Undergraduate Texts in Mathematics

George E. Martin

Transformation Geometry
An Introduction to Symmetry

With 221 Illustrations

Springer-Verlag
New York   Heidelberg   Berlin

George E. Martin
Department of Mathematics and Statistics
State University of New York at Albany
1400 Washington Avenue
Albany, NY 12222
U.S.A.

AMS Subject Classification (1980): 51-01, 51A25, 51A40, 05-01, 05B25, 05B30

Library of Congress Cataloging in Publication Data

Martin, George Edward, 1932–
  Transformation geometry.

  (Undergraduate texts in mathematics)
  Bibliography: p.
  Includes indexes.
  1. Transformations (Mathematics) 2. Geometry.
3. Symmetry. I. Title. II. Series.
QA601.M36 1982   516′.1   82-756
                                     AACR2

Printed in the United States of America.

9 8 7 6 5 4 3 2 1

ISBN 0-387-90636-3 Springer-Verlag New York Heidelberg Berlin
ISBN 3-540-90636-3 Springer-Verlag Berlin Heidelberg New York

To Margaret

Preface

Transformation geometry is a relatively recent expression of the successful venture of bringing together geometry and algebra. The name describes an approach as much as the content. Our subject is Euclidean geometry. Essential to the study of the plane or any mathematical system is an understanding of the transformations on that system that preserve designated features of the system.

Our study of the automorphisms of the plane and of space is based on only the most elementary high-school geometry. In particular, group theory is *not* a prerequisite here. On the contrary, this modern approach to Euclidean geometry gives the concrete examples that are necessary to appreciate an introduction to group theory. Therefore, a course based on this text is an excellent prerequisite to the standard course in abstract algebra taken by every undergraduate mathematics major. An advantage of having no college mathematics prerequisite to our study is that the text is then useful for graduate mathematics courses designed for secondary teachers. Many of the students in these classes either have never taken linear algebra or else have taken it too long ago to recall even the basic ideas. It turns out that very little is lost here by not assuming linear algebra. A preliminary version of the text was written for and used in two courses—one was a graduate course for teachers and the other a sophomore course designed for the prospective teacher and the general mathematics major taking one course in geometry.

A several-track option allows the basic material on isometries to be followed by application of isometries to the ornamental groups, tessellations, similarities and their application to classical theorems, affine transformations, and transformations on three-space. The illustration at the end of this Preface shows the interdependence among the chapters. The book may be used for either a one-semester or a one-year course. The first nine chapters are necessary and sufficient for such a course.

Instructors who pursue the study of the ornamental groups in Chapter 12 will want to know about the *Dover Pictorial Archive Series* by Dover Publications of New York. Several of the illustrations in this text are taken from *Designs and Patterns from Historical Ornament* by W. and G. Audsley and other paperback books in that series of copyright-free art. Those feeling strongly about including or excluding the traditional topics that constitute what was called *college geometry* a few years ago may choose to emphasize or omit Chapter 14. Such decisions permit a course with the instructor's own mark. Perhaps some students will use chapters not covered in class for independent study.

The belief that geometries can be classified by their symmetry groups is no longer tenable. However, the correspondence for the classical geometries and their groups remains valid. Undergraduates should not be expected to grasp the idea of Klein's Erlanger program before encountering at least the projective and hyperbolic geometries. Therefore, although the basic spirit of the text is to begin to carry out Klein's program, little mention of the program is made within the text.

I am indebted to my colleagues Hugh Gordon and Violet Larney for classroom testing the preliminary version of the text at the undergraduate and graduate levels. I am especially indebted to Hugh for the discussions and many helpful suggestions offered at our lunch meetings over the years. Finally, I would like to thank Anne Marie Vancura for producing the art work.

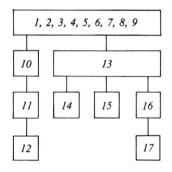

George E. Martin

Contents

Chapter 17
Space and Symmetry

Chapter 1

Introduction

§1.1 Transformations and Collineations

There are only seventeen mathematically different types of wallpaper patterns. This intriguing result, which was proved in 1891 and which we shall encounter as an application of our study of transformations, was empirically known to the Moors of fifteenth-century Spain but was unknown to the Greek mathematician Euclid. Euclid's *Elements* first placed mathematics on an axiomatic basis and was probably written in Alexandria shortly after 300 B.C. Euclid also told King Ptolemy there was no royal road to geometry. From the first few of Euclid's propositions it is evident that, were he alive today, Euclid would present much of his mathematics through transformations. *Transformation geometry* is a stretch along the royal road to geometry. The name describes an approach as much as the content of our study. Our subject is Euclidean geometry and, until Chapter 16 where we consider three-space, we suppose we are talking about the Euclidean plane.

A *transformation* on the plane is a one-to-one correspondence from the set of points in the plane onto itself. For a given transformation f, this means that for every point P there is a unique point Q such that $f(P) = Q$ and, conversely, for every point R there is a unique point S such that $f(S) = R$. To fully understand any mathematical system you must understand the transformations of the system and especially those transformations of the system that leave some particular aspect of the system invariant. The set of all transformations on the plane is too large to be very interesting. One very large subset consists of those transformations that preserve *nearness*, meaning point P near point Q implies $f(P)$ near point $f(Q)$. Making this meaning precise and subsequent study fall into the mathematical realm called *topology*. Topology has been called *rubber sheet geometry* since a line

1

need not go into a line under such a transformation. However, we shall study the more classical transformations f having the property that if l is a line then $f(l)$ is also a line. A transformation f with this property is called a **collineation**. We take the view that a line is a set of points and $f(l)$ is the set of all points $f(P)$ with point P on line l. For a given collineation f, then $f(l)$ is a line and point $f(P)$ is on line $f(l)$ if and only if point P is on line l. Translations and rotations are examples of collineations. Our study might well be called "collineation geometry." In any case, let us realize that we shall be studying Euclidean geometry using the methods supplied by emphasizing transformations.

To say that for every point P there is a point Q such that $f(P) = Q$ is to say that f is a **mapping** on the set of points. Mapping f is said to be **onto** if for every point P there is a point Q such that $f(Q) = P$; mapping f is said to be **one-to-one** if $f(R) = f(S)$ implies $R = S$. So a transformation is just a mapping on the points that is both one-to-one and onto. (*Transformation* may be defined differently in other courses, such as linear algebra.)

Points will be denoted by capital Roman letters. Lines will be denoted by lowercase Roman letters. Transformations will be denoted by lowercase Greek letters. The reason for this last choice is to make expressions easier to read, once you become familiar with the letters. You already know alpha α, beta β, and gamma γ from trigonometry and perhaps also theta Θ, phi Φ, and psi Ψ. Four more letters that we shall use often are rho ρ, sigma σ, tau τ, and iota ι. (Learn to pronounce the names of the letters the way your instructors do—even if this means different pronunciations in different classes.) For reference the entire Greek alphabet is given below.

The Greek Alphabet

Letters		Names	Letters		Names	Letters		Names
A	α	alpha	I	ι	iota	P	ρ	rho
B	β	beta	K	κ	kappa	Σ	σ	sigma
Γ	γ	gamma	Λ	λ	lambda	T	τ	tau
Δ	δ	delta	M	μ	mu	Υ	υ	upsilon
E	ε	epsilon	N	ν	nu	Φ	ϕ	phi
Z	ζ	zeta	Ξ	ξ	xi	X	χ	chi
H	η	eta	O	o	omicron	Ψ	ψ	psi
Θ	θ	theta	Π	π	pi	Ω	ω	omega

We suppose familiarity with the **Cartesian plane**. Here, a **point** is an ordered pair (x, y) of numbers, a **line** is the set of points satisfying an equation $aX + bY + c = 0$ where a, b, c are numbers with not both $a = 0$ and $b = 0$, and the **distance** between points (x_1, y_1) and (x_2, y_2) is given by

$$\sqrt{(x_2 - x_1)^2 + (y_2 - y_1)^2}.$$

Note that numbers and lines are both denoted by lower case Roman letters; there is little likelihood of confusion, however. Let's agree once and for all that O is the *origin* $(0, 0)$ in the context of the Cartesian plane. The Cartesian plane is the standard model of the Euclidean plane, and, for our purposes, the two may be identified. We shall use "Cartesian plane" to emphasize that we are dealing with the analytic aspect of the plane. The name comes from that of Rene Descartes (1596–1650), who is usually given credit for the invention of analytic geometry and is often called the founder of modern philosophy. There is very little of importance in geometry between Pappus' *Collection*, which was written in Alexandria about A.D. 320, and Descartes' *The Geometry*, published in 1637. Descartes' bringing together of geometry and algebra released the bind which essentially limited the advancement of Greek mathematics. Transformation geometry is a more recent expression of the successful venture of bringing geometry and algebra together.

We can use the Cartesian plane to provide some examples. The mapping α on the Cartesian plane that sends (x, y) to (x^2, y) is not a transformation since there is no point (x, y) such that $\alpha((x, y)) = (-1, 2)$. The mapping β on the plane that sends (x, y) to (x, y^3) is a transformation as $(u, v^{1/3})$ is the unique point sent to (u, v) for given numbers u and v. However, β is not a collineation since the line with equation $Y = X$ is not sent to a line but rather to the cubic curve with equation $Y = X^3$. Let us now show that the mapping γ that sends each point (x, y) to the point $(-x + y/2, x + 2)$ is a collineation. Setting $u = -x + y/2$ and $v = x + 2$, we have unique solutions $x = v - 2$ and $y = 2u + 2v - 4$ such that $\gamma((x, y)) = (u, v)$ for any numbers u and v. Hence, γ is a transformation. Then, the following are equivalent where $\alpha((x, y)) = (u, v)$:

(1) (x, y) on line with equation $aX + bY + c = 0$.
(2) $ax + by + c = 0$.
(3) $a(v - 2) + b(2u + 2v - 4) + c = 0$.
(4) $(2b)u + (a + 2b)v + (c - 4b - 2a) = 0$.
(5) (u, v) on line with equation $(2b)X + (a + 2b)Y + (c - 4b - 2a) = 0$.

So the line with equation $a_1 X + b_1 Y + c_1 = 0$ goes to the line with equation $a_2 X + b_2 Y + c_2 = 0$ where $a_2 = 2b_1$, $b_2 = a_1 + 2b_1$, and $c_2 = c_1 - 4b_1 - 2a_1$. Hence, γ is a collineation.

§1.2 Geometric Notation

You must master the notation used in the text so that you can understand what it is you are reading. The notation in the next paragraph is basic. The remaining material is all used somewhere later and should be familiar. This should be read now to emphasize the basic notation and used later as a reference.

Number AB is the *distance* from point A to point B. By the *triangle inequality*, $AB + BC \geq AC$ for points A, B, C. "A–B–C" is read "point B is between points A and C" and means A, B, C are three distinct points such that $AB + BC = AC$. $\overleftrightarrow{AB}$ is the unique *line* determined by two points A and B. (Read "$\overleftrightarrow{AB}$" as "line $A B$" and not as "$A B$" since $\overleftrightarrow{AB} \neq AB$.) $\overline{AB}$ is a *segment* and consists of A, B, and all points between A and B. $\overrightarrow{AB}$ is a *ray* and consists of all points in $\overline{AB}$ together with all points P such that A–B–P. $\angle ABC$ is an *angle* and is the union of noncollinear rays $\overrightarrow{BA}$ and $\overrightarrow{BC}$. $m\angle ABC$ is the degree measure of $\angle ABC$ and is a number between 0 and 180. $\triangle ABC$ is a *triangle* and is the union of noncollinear segments $\overline{AB}, \overline{BC}$, and $\overline{CA}$.

You probably know that "iff" means "if and only if." Also, "off" means "not on," as in "the line is off the point."

"$\cong$" is read "is congruent to" and has various meanings depending on context. $\overline{AB} \cong \overline{CD}$ iff $AB = CD$. $\angle ABC \cong \angle DEF$ iff $m\angle ABC = m\angle DEF$. $\triangle ABC \cong \triangle DEF$ iff $\overline{AB} \cong \overline{DE}, \overline{BC} \cong \overline{EF}, \overline{AC} \cong \overline{DF}, \angle A \cong \angle D, \angle B \cong \angle E$, and $\angle C \cong \angle F$. Not all six corresponding parts must be checked to show triangles congruent. The familiar congruence theorems for triangles $\triangle ABC$ and $\triangle DEF$ are:

(1) SAS: If $\overline{AB} \cong \overline{DE}, \angle A \cong \angle D$, and $\overline{AC} \cong \overline{DF}$, then $\triangle ABC \cong \triangle DEF$.
(2) ASA: If $\angle A \cong \angle D, \overline{AB} \cong \overline{DE}$, and $\angle B \cong \angle E$, then $\triangle ABC \cong \triangle DEF$.
(3) SAA: If $\overline{AB} \cong \overline{DE}, \angle B \cong \angle E$, and $\angle C \cong \angle F$, then $\triangle ABC \cong \triangle DEF$.
(4) SSS: If $\overline{AB} \cong \overline{DE}, \overline{BC} \cong \overline{EF}$, and $\overline{CA} \cong \overline{FD}$, then $\triangle ABC \cong \triangle DEF$.

The *Exterior Angle Theorem* states that given $\triangle ABC$ and B–C–D, then $m\angle ACD = m\angle A + m\angle B$. So for $\triangle ABC$ we have $m\angle A + m\angle B + m\angle C = 180$. Given $\triangle ABC$ and $\triangle DEF$ such that $\angle A \cong \angle D, \angle B \cong \angle E$, and $\angle C \cong \angle F$, then $\triangle ABC \sim \triangle DEF$, where "$\sim$" is read "is similar to." If two of these three angle congruences hold then the third congruence necessarily holds and the triangles are similar; this result is known as the *Angle–Angle Similarity Theorem*. Two triangles are also similar iff their corresponding sides are proportional.

We suppose familiarity with *directed angles* and *directed angle measure*, say from $\overrightarrow{AB}$ to $\overrightarrow{AC}$, with counterclockwise orientation chosen as positive, and clockwise orientation chosen as negative. In general, for real numbers Θ and Φ we agree that $\Theta° = \Phi°$ iff $\Theta = \Phi + k(360)$ for some integer k.

Given line l, the points of the plane are partitioned into three sets, namely the line itself and the two *halfplanes* or *sides* of the line. *Pasch's Axiom* requires that a line l intersecting $\triangle ABC$ at a point between A and B must intersect the triangle in at least one more point (which is unique unless $l = \overleftrightarrow{AB}$). A set of points is *convex* if for any two points A and B in the set all the points between A and B are in the set.

Lines l and m in the plane are *parallel* iff either $l = m$ or else l and m have no point in common. In planar coordinate geometry, point (x, y) is on the line with equation $aX + bY + c = 0$ iff $ax + by + c = 0$. So point (x, y)

is off the line with equation $aX + bY + c = 0$ iff $ax + by + c \neq 0$. Lines with equations $aX + bY + c = 0$ and $dX + eY + f = 0$ are parallel iff $ae - bd = 0$ and are perpendicular iff $ad + be = 0$.

In three-dimensional space, lines l and m are *parallel* iff either $l = m$ or else l and m are coplanar lines that do not intersect. Nonintersecting lines that are not coplanar are *skew* to each other. Planes Γ and Δ are *parallel* iff either $\Gamma = \Delta$ or else Γ and Δ do not intersect. Line l and plane Π are *parallel* iff either l is on Π or else l and Π do not intersect.

In the plane, the locus of all points equidistant from two points A and B is the perpendicular bisector of A and B, which is a line through the midpoint of $\overline{AB}$ and perpendicular to $\overline{AB}$. In space, the locus of all points equidistant from two points A and B is the perpendicular bisector of A and B, which is a plane through the midpoint of $\overline{AB}$ and perpendicular to $\overline{AB}$.

§1.3 Exercises

1.1. Which of the mappings defined on the Cartesian plane by the equations below are transformations?

$$\alpha((x, y)) = (x^3, y^3), \qquad \beta((x, y)) = (\cos x, \sin y), \qquad \gamma((x, y)) = (x^3 - x, y),$$

$$\delta((x, y)) = (2x, 3y), \qquad \varepsilon((x, y)) = (-x, x + 3), \qquad \eta((x, y)) = (3y, x + 2),$$

$$\rho((x, y)) = (\sqrt[3]{x}, e^y), \qquad \sigma((x, y)) = (-x, -y), \qquad \tau((x, y)) = (x + 2, y - 3).$$

1.2. Which of the transformations in the exercise above are collineations? For each collineation, find the image of the line with equation $aX + bY + c = 0$.

1.3. Without looking back in the text, write in your own words definitions for *transformation* and *collineation*. Then compare to see whether your definitions are equivalent to those in the text.

1.4. Find the image of the line with equation $Y = 5X + 7$ under collineation α if $\alpha((x, y))$ is:
   (a) $(-x, y)$,   (b) $(x, -y)$,   (c) $(-x, -y)$,   (d) $(2y - x, x - 2)$.

1.5. True or False
   Suppose σ is a transformation on the plane.
   (a) If $\sigma(P) = \sigma(Q)$, then $P = Q$.
   (b) For any point P there is a unique point Q such that $\sigma(P) = Q$.
   (c) For any point P there is a point Q such that $\sigma(P) = Q$.
   (d) For any point P there is a unique point Q such that $\sigma(Q) = P$.
   (e) For any point P there is a point Q such that $\sigma(Q) = P$.
   (f) A collineation is necessarily a transformation.
   (g) A transformation is necessarily a collineation.
   (h) A collineation is a mapping that is one-to-one.
   (i) A collineation is a mapping that is onto.
   (j) A transformation is onto but not necessarily one-to-one.

1.6. Let $M = (a, b)$ ın the Cartesian plane. Find equations for x' and y' where $\alpha((x, y)) = (x', y')$ and α is the mapping such that for any point P the midpoint of P and $\alpha(P)$ is always M.

1.7. Copy each of the lowercase Greek letters five times or else use the Greek letters to make up some "sillies" such as [Φ, Φ, foe fum] and [Φ], where the first is a familiar line that uses the two common ways of pronouncing phi and the second denotes a good phonograph.

1.8. Learn the Greek alphabet.

1.9. Give three examples of transformations on the Cartesian plane that are not collineations.

1.10. Fill in the table with appropriate "Yes" or "No" for mapping α on the Cartesian plane.

$\alpha((x, y))$	$(\lvert x \rvert, \lvert y \rvert)$	(e^x, e^y)	$(x^3 - x^2, y)$	(x^3, y^3)	$(y/2, 2x)$	$(x/2, y/2)$	$(x - 2, 3y)$	$(-y, -x)$
α is 1–1								
α is onto								
α is a transformation								
α is a collineation								

1.11. Show that a collineation determines a one-to-one correspondence from the set of all lines onto itself.

1.12. Produce two noncongruent triangles such that five parts of one are congruent to five parts of the other.

1.13. Find the preimage of the line with equation $Y = 3X + 2$ under the collineation α where $\alpha((x, y)) = (3y, x - y)$.

1.14. Show the lines with equations $aX + bY + c = 0$ and $dX + eY + f = 0$ are parallel iff $ae - bd = 0$ and are perpendicular iff $ad + be = 0$.

1.15. Prove or disprove: A mapping on the Cartesian plane that preserves betweenness among the points is necessarily a collineation.

Chapter 2

Properties of Transformations

§2.1 Groups of Transformations

The *identity* transformation ι is defined by $\iota(P) = P$ for every point P. No other transformation is allowed to use this Greek letter iota. As you can see, ι is in some sense actually the least exciting of all the transformations. If ι is in set $\mathscr{G}$ of transformations, then $\mathscr{G}$ is said to have the *identity property*. We continue below to look at properties of a set $\mathscr{G}$ of transformations that make $\mathscr{G}$ algebraically interesting.

Recall that γ is a transformation iff for every point P there is a unique point Q such that $\gamma(P) = Q$ and, conversely, for every point P there is a unique point Q such that $\gamma(Q) = P$. From this definition we see that γ^{-1} is also a transformation where γ^{-1} is the mapping defined by $\gamma^{-1}(A) = B$ iff $A = \gamma(B)$. The transformation γ^{-1} is called the *inverse* of transformation γ. We read "γ^{-1}" as "gamma inverse." If γ^{-1} is also in $\mathscr{G}$ for every transformation γ in our set $\mathscr{G}$ of transformations, then $\mathscr{G}$ is said to have the *inverse property*.

Whenever two transformations are brought together they might form new transformations. In fact, one transformation might form new transformations by itself, as we can see by considering $\alpha = \beta$ below. The *composite* $\beta \circ \alpha$ of transformations α and β is the mapping defined by $\beta \circ \alpha(P) = \beta(\alpha(P))$ for every point P. Note that α is applied first and then β is applied. We read "$\beta \circ \alpha$" as "beta following alpha." Since for every point C there is a point B such that $\beta(B) = C$ and for every point B there is a point A such that $\alpha(A) = B$, then for every point C there is a point A such that $\beta \circ \alpha(A) = \beta(\alpha(A)) = \beta(B) = C$. So $\beta \circ \alpha$ is an onto mapping. Also, $\beta \circ \alpha$ is one-to-one as the following argument shows. Suppose $\beta \circ \alpha(P) = \beta \circ \alpha(Q)$. Then $\beta(\alpha(P)) = \beta(\alpha(Q))$ by the definition of composite $\beta \circ \alpha$. So $\alpha(P) = \alpha(Q)$ since β is one-to-one. Then $P = Q$ as α is one-to-one. Therefore, $\beta \circ \alpha$ is both one-to-one and onto. We have thus proved the following theorem.

Theorem 2.1. *The composite $\beta \circ \alpha$ of transformations α and β is itself a transformation.*

If our set $\mathscr{G}$ has the property that the composite $\beta \circ \alpha$ is in $\mathscr{G}$ whenever α and β are in $\mathscr{G}$, then $\mathscr{G}$ is said to have the **closure property**. Since both $\gamma^{-1} \circ \gamma(P) = P$ and $\gamma \circ \gamma^{-1}(P) = P$ for every point P, we see that $\gamma^{-1} \circ \gamma = \gamma \circ \gamma^{-1} = \iota$ for every transformation γ. Hence, if $\mathscr{G}$ is a nonempty set of transformations having both the inverse property and the closure property, then $\mathscr{G}$ must necessarily have the identity property.

Our set $\mathscr{G}$ of transformations is said to have the **associative property**, as any elements α, β, γ in $\mathscr{G}$ satisfy the **associative law**: $\gamma \circ (\beta \circ \alpha) = (\gamma \circ \beta) \circ \alpha$. The proof of this is given below, except that as Exercise 2.1 you will be asked to give the reason for each equality sign. For every point P,

$$\begin{aligned}
[\gamma \circ (\beta \circ \alpha)](P) &= \gamma((\beta \circ \alpha)(P)) \\
&= \gamma(\beta(\alpha(P))) \\
&= (\gamma \circ \beta)(\alpha(P)) \\
&= [(\gamma \circ \beta) \circ \alpha](P).
\end{aligned}$$

The important sets of transformations are those that simultaneously satisfy the closure property, the associative property, the identity property, and the inverse property. Such a set is said to form a **group**. We mention all four properties because it is these four properties that are used for the generalized definition of a *group* in abstract algebra. However, when we want to check that a nonempty set $\mathscr{G}$ of transformations forms a group, we need check only the closure property and the inverse property. Since these two properties hold for the set of all transformations, we have the first part of the following theorem.

Theorem 2.2. *The set of all transformations forms a group. The set of all collineations forms a group.*

For the proof of the second part of the theorem, we suppose α and β are collineations. Suppose l is a line. Then $\alpha(l)$ is a line since α is a collineation, and $\beta(\alpha(l))$ is then a line since β is a collineation. Hence, $\beta \circ \alpha(l)$ is a line, and $\beta \circ \alpha$ is a collineation. So the set of collineations satisfies the closure property. There is a line m such that $\alpha(m) = l$. (Why?) So $\alpha^{-1}(l) = \alpha^{-1}(\alpha(m)) = \alpha^{-1} \circ \alpha(m) = \iota(m) = m$. Hence α^{-1} is a collineation, and the set of all collineations satisfies the inverse property. The set is not empty as the identity is a collineation. Therefore, the set of all collineations forms a group.

If every element of transformation group $\mathscr{G}_2$ is an element of transformation group $\mathscr{G}_1$, then $\mathscr{G}_2$ is a **subgroup** of $\mathscr{G}_1$. All of our groups will be subgroups of the group of all collineations. These transformation groups will be a very important part of our study. There will be more examples of groups in the next section and throughout the remainder of this book.

We should take note that the word *group* now has a technical meaning and should never be used as a general collective noun in place of the word *set*. The technical word was first used by the eccentric French Republican Evariste Galois (1811–1832). After failing an entrance examination to the Ecole Polytechnique, the seventeen-year-old Galois worked up his mathematical discoveries into a paper for the mathematician Cauchy to present to the Académie. Cauchy lost the paper. A second entrance examination resulted in failure, and a second paper submitted to the mathematician Fourier was lost. Galois then submitted what is one of the most beautiful pieces of modern mathematics, which is now called Galois theory, to the Académie. This was rejected as incomprehensible by the mathematician Poisson. Galois joined the National Guard. The night before an absurd duel he spent hours jotting down the outlines of his research. The duel was fatal for the twenty-one-year-old mathematical genius, who had in his haste, the night before, written in the margin, "I have no time."

Transformations α and β may or may not satisfy the **commutative law**: $\alpha \circ \beta = \beta \circ \alpha$. If the commutative law is always satisfied by the elements from a group, then that group is said to be **abelian** or **commutative**. The term *abelian* is after the brilliant Norwegian mathematician Neils Henrik Abel (1802–1829), who died of tuberculosis at twenty-six. Cauchy, who you recall mislaid the work of Galois, also mislaid Abel's paper on elliptic functions. This paper was later called, "a monument more lasting than bronze." Abel is best known for disposing of the long-outstanding problem of solving the quintic equation. In 1824, Abel showed that a general solution is impossible: There is no general formula expressed in algebraic operations on the coefficients of a polynomial equation that gives a solution to the equation if the degree of the polynomial equation is greater than 4.

§2.2 Involutions

Life becomes somewhat easier once we admit that we are basically lazy. We rationalize this by saying we find it *convenient* to streamline our notation. Thus we abbreviate "$\beta \circ \alpha$" as "$\beta\alpha$" and talk about "the product of α multiplied by β on the left" or "the product of β multiplied by α on the right" or just "the product beta-alpha." We claim no confusion will arise if we drop the symbol denoting composition. Since the associative law holds for composition of transformations, then "$(\varepsilon \circ (\delta \circ \gamma)) \circ (\beta \circ \alpha)$" is now written "$\varepsilon\delta\gamma\beta\alpha$." In "$\gamma = \alpha_n\alpha_{n-1} \cdots \alpha_2\alpha_1$" with n a positive integer, transformation γ is expressed as a **product of n transformations**. Also, if all the α_i are equal to α, then we write "$\gamma = \alpha^n$." Further, if $\alpha = \beta^{-1}$, then we write "$\gamma = \beta^{-n}$."

As expected, we define β^0 to be ι for any transformation β. So $(\beta^{-1})^n = \beta^{-n}$ for any transformation β and any integer n.

In Exercise 2.2 you will be asked to prove the next theorem. The first implication in the theorem is called the **right cancellation law**; the second is called the **left cancellation law**.

Theorem 2.3. *If α, β, and γ are elements in a group, then $\beta\alpha = \gamma\alpha$ implies $\beta = \gamma$, $\beta\alpha = \beta\gamma$ implies $\alpha = \gamma$, $\beta\alpha = \alpha$ implies $\beta = \iota$, $\beta\alpha = \beta$ implies $\alpha = \iota$, and $\beta\alpha = \iota$ implies $\beta = \alpha^{-1}$ and $\alpha = \beta^{-1}$.*

From the easily verified equations

$$(\omega \cdots \gamma\beta\alpha)(\alpha^{-1}\beta^{-1}\gamma^{-1} \cdots \omega^{-1}) = \iota,$$

$$(\alpha^{-1}\beta^{-1}\gamma^{-1} \cdots \omega^{-1})(\omega \cdots \gamma\beta\alpha) = \iota,$$

we see that $(\omega \cdots \gamma\beta\alpha)^{-1} = \alpha^{-1}\beta^{-1}\gamma^{-1} \cdots \omega^{-1}$. This result is stated as our next theorem.

Theorem 2.4. *In a group, the inverse of a product is the product of the inverses in reverse order.*

If group $\mathscr{G}$ has exactly n elements, then $\mathscr{G}$ is said to be **finite** and have **order** n; otherwise $\mathscr{G}$ is said to be **infinite**. Analogously, if there is a smallest positive integer n such that $\alpha^n = \iota$, then transformation α is said to have **order** n; otherwise α is said to have **infinite order**. For example, let ρ be a rotation of $360/n$ degrees about the origin with n a positive integer and let $\tau((x, y)) = (x + 1, y)$. Then ρ has order n, the set $\{\rho, \rho^2, \ldots, \rho^n\}$ forms a group of order n, element τ has infinite order, and the set of all transformations τ^k with k an integer forms an infinite group. In Exercise 2.6 you are asked to show in general that if transformation α has order n then $\{\alpha, \alpha^2, \ldots, \alpha^n\}$ forms a group of order n, while if α has infinite order then the set of all integral powers of α forms an infinite group.

If every element of a group containing α is a power of α, then we say that the group is **cyclic** with **generator** α and denote the group as $\langle\alpha\rangle$. For example, if ρ is a rotation of $36°$, then $\langle\rho\rangle$ is a cyclic group of order 10. Note that this same group is generated by β where $\beta = \rho^3$. In fact, we have $\langle\rho\rangle = \langle\rho^3\rangle = \langle\rho^7\rangle = \langle\rho^9\rangle$. So a cyclic group may have more than one generator. Since the powers of a transformation always commute, i.e., $\alpha^n\alpha^m = \alpha^{n+m} = \alpha^{m+n} = \alpha^m\alpha^n$ for integers m and n, we see that a cyclic group is always abelian.

The notation for a cyclic group introduced above is generalized as follows. If $\mathscr{G} = \langle\alpha, \beta, \gamma, \ldots\rangle$, then every element of group $\mathscr{G}$ can be written as a product of powers of α, β, γ, $\ldots$ and $\mathscr{G}$ is said to be **generated** by $\{\alpha, \beta, \gamma, \ldots\}$. Such a product might be $\alpha^5\beta^3\alpha^{-2}\gamma^{-7}\beta^4$.

Foremost among the particular transformations that will command our attention are the **involutions**, which are the transformations of order 2. In other words, transformation γ is an involution iff $\gamma^2 = \iota$ but $\gamma \neq \iota$. For still

Table 2.1

C_4	ι	ρ	ρ^2	ρ^3
ι	ι	ρ	ρ^2	ρ^3
ρ	ρ	ρ^2	ρ^3	ι
ρ^2	ρ^2	ρ^3	ι	ρ
ρ^3	ρ^3	ι	ρ	ρ^2

Table 2.2

V_4	ι	σ_h	σ_v	σ_O
ι	ι	σ_h	σ_v	σ_O
σ_h	σ_h	ι	σ_O	σ_v
σ_v	σ_v	σ_O	ι	σ_h
σ_O	σ_O	σ_v	σ_h	ι

another characterization, we see that nonidentity transformation γ is an involution iff $\gamma = \gamma^{-1}$. This last observation follows from the fact that equations $\gamma^2 = \iota$ and $\gamma = \gamma^{-1}$ are easily seen to be equivalent by multiplying both sides of the first by γ^{-1} and by multiplying both sides of the second by γ. Note that the identity is not an involution by definition.

A multiplication table for a finite group is often called a *Cayley table* for the group. This is in honor of the mathematically prolific algebraist Arthur Cayley, who was one of the first to study matrices and who is credited with introducing the ordinary analytic geometry of n-dimensional space. In a Cayley table, the product $\beta\alpha$ is found in the row headed "β" and the column headed "α." Table 2.1 gives a Cayley table for the group C_4 that is generated by a rotation ρ of 90° about the origin. Table 2.2 gives the Cayley table for the group V_4, where the elements of V_4 are defined on the Cartesian plane by

$$\iota((x, y)) = (x, y), \qquad \sigma_O((x, y)) = (-x, -y),$$

$$\sigma_h((x, y)) = (x, -y), \qquad \sigma_v((x, y)) = (-x, y).$$

From their Cayley tables, it is easy to check the closure property and the inverse property for each of C_4 and V_4. So C_4 and V_4 are certainly groups and each has order 4. Group C_4 is cyclic and is generated by ρ. Since $(\rho^3)^2 = \rho^6 = \rho^2$, $(\rho^3)^3 = \rho^9 = \rho$, and $(\rho^3)^4 = \rho^{12} = \iota$, then C_4 is also generated by ρ^3. So $C_4 = \langle\rho\rangle = \langle\rho^3\rangle$. Group C_4 contains the one involution ρ^2. Group V_4 is abelian but not cyclic. Every element of V_4 except the identity is an involution. Evidently, $V_4 = \langle\sigma_h, \sigma_v\rangle = \langle\sigma_h, \sigma_O\rangle = \langle\sigma_v, \sigma_O\rangle$.

The Cayley table for V_4 can be computed algebraically without any geometric interpretation. For example, since $\sigma_h\sigma_O((x, y)) = \sigma_h((-x, -y)) = (-x, y)$ for all (x, y), then $\sigma_h\sigma_O = \sigma_v$. However, you probably recall that σ_h is just the reflection in the X-axis, σ_v is the reflection in the Y-axis, and σ_O is the rotation about the origin of 180°. So ρ^2 and σ_O are the same transformation.

You will have noticed that Chapters 1 and 2 have been concerned mostly with vocabulary. Armed with this vocabulary, we next begin our study of specific groups of transformations. However, we might pause to emphasize the often-overlooked fact that definitions are just as important in mathematics as the theorems. You should be able to define each of the words or phrases that appears in boldface italics as well as the symbols. If you are going to memorize anything, then memorize the definitions. Remember that if you don't know the meaning of the words you use then you don't know what you're talking about. Can you define the following ten key terms that have been introduced in the first two chapters: *transformation, one-to-one, onto, composite, collineation, Cartesian plane, group, order, involution, generator?*

§2.3 Exercises

2.1. Verify that every set of transformations has the associative property.

2.2. Prove Theorem 2.3.

2.3. Show that in a Cayley table for a finite group, each element of the group appears exactly once in each row and exactly once in each column.

2.4. Find the image of each of $(2, 3), (-2, -3), (-2, 3), (2, -3)$ under each of the involutions in V_4. Also, verify the Cayley table for V_4 by the algebraic methods described above.

2.5. True or False
   (a) If α and β are transformations, then $\alpha = \beta$ iff $\alpha(P) = \beta(P)$ for every point P.
   (b) Transformation ι is in every group of transformations.
   (c) If $\alpha\beta = \iota$, then $\alpha = \beta^{-1}$ and $\beta = \alpha^{-1}$ for transformations α and β.
   (d) "$\alpha \circ \beta$" is read "beta following alpha."
   (e) If α and β are both in group $\mathcal{G}$, then $\alpha\beta = \beta\alpha$.
   (f) $\langle \iota \rangle$ is a cyclic group of order 1.
   (g) $\langle \gamma \rangle = \langle \gamma^{-1} \rangle$ for any transformation γ.
   (h) An abelian group is always cyclic, but a cyclic group is not always abelian.
   (i) $(\alpha\beta)^{-1} = \alpha^{-1}\beta^{-1}$ for transformations α and β.
   (j) If $\langle \alpha \rangle = \langle \beta \rangle$, then $\alpha = \beta$ or $\alpha = \beta^{-1}$.

2.6. Prove: If transformation α has order n, then $\{\alpha, \alpha^2, \alpha^3, \ldots, \alpha^n\}$ forms a group of order n; if transformation α has infinite order, then the set of all integral powers of α forms an infinite group.

2.7. Prove or disprove: There is an infinite cyclic group of rotations.

2.8. Prove or disprove: A nonidentity cyclic group has at least two generators.

2.9. Prove or disprove: Every group that contains at least two elements contains an involution.

2.10. Prove or disprove: Any group of order 4 has a Cayley table that except for notation denoting the elements is like that of C_4 or V_4.

2.11. Read one of the biographical chapters on Descartes, Abel, Galois, or Cayley from *Men of Mathematics* by E. T. Bell.

2.12. Find out more about Galois by looking in the book *Whom the Gods Love* by Leopold Infeld.

2.13. Find all a and b such that α is an involution if $\alpha((x, y)) = (ay, x/b)$.

2.14. Divide a given rectangular region into two congruent regions in five different ways.

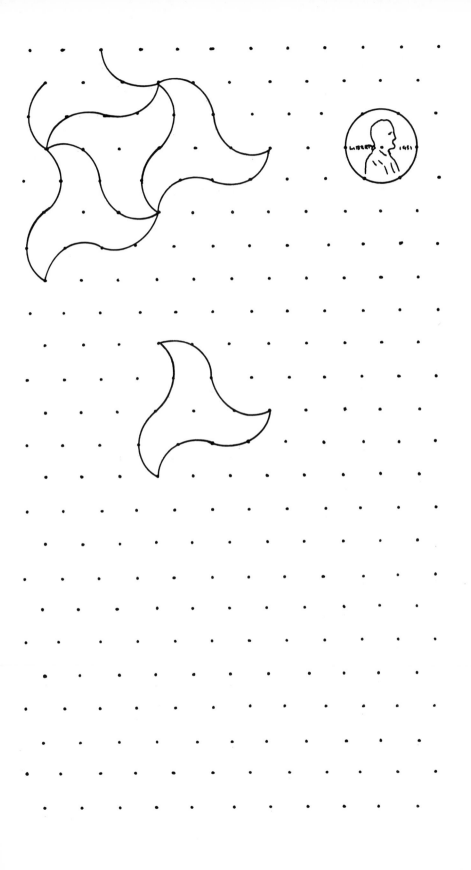

Chapter 3

Translations and Halfturns

§3.1 Translations

The methods of this chapter are from *analytic geometry*. To emphasize the use of numbers, we often call this approach *coordinate geometry*. In either case, we start with the *translations* as they are important and so easily studied by these methods. Later, we shall not hesitate to use the older synthetic methods from the geometry of Euclid when they are most convenient. Then, as we progress, we shall be using more and more group-algebraic methods. At times it may even be hard to say whether a particular part is essentially synthetic, analytic, or algebraic, but this will be completely unimportant.

When we say *mapping* α *has equations*

$$\begin{cases} x' = ax + by + c, \\ y' = dx + ey + f, \end{cases}$$

we mean $(x', y') = \alpha((x, y))$ for each point (x, y) in the Cartesian plane, where a, b, c, d, e, f are numbers. A *translation* is a mapping having equations of the form

$$\begin{cases} x' = x + a, \\ y' = y + b. \end{cases}$$

Given any two of (x, y), (x', y'), and (a, b), the third is then uniquely determined by this last set of equations. Hence, not only is a translation a transformation, but there is a unique translation taking given point P to given point Q. For example, let $P = (c, d)$ and $Q = (e, f)$. Then there are unique

14

numbers a and b such that $e = c + a$ and $f = d + b$. So the unique translation that takes P to Q has equations

$$x' = x + (e - c) \quad \text{and} \quad y' = y + (f - d).$$

We shall denote this unique translation by $\tau_{P,Q}$ and use the Greek letter tau only for translations. We have proved the following.

Theorem 3.1. *Given points P and Q, there is a unique translation taking P to Q, namely $\tau_{P,Q}$.*

By the theorem above, if $\tau_{P,Q}(R) = S$, then $\tau_{P,Q} = \tau_{R,S}$ for points P, Q, R, S. Note that the identity is a special case of a translation as $\iota = \tau_{P,P}$ for each point P. Also, if $\tau_{P,Q}(R) = R$ for point R, then $P = Q$ as $\tau_{P,Q} = \tau_{R,R} = \iota$.

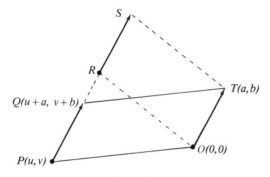

Figure 3.1

Now let $T = (a, b)$ and $P = (u, v)$. Then $\tau_{O,T}$ is the translation taking $(0, 0)$ to (a, b) and has equations $x' = x + a$ and $y' = y + b$. So $\tau_{O,T}((x, y)) = (x + a, y + b)$. Suppose $Q = \tau_{O,T}(P)$. See Figure 3.1. Then $\tau_{O,T} = \tau_{P,Q}$ and $Q = (u + a, v + b)$. It is easy to check that $OT = PQ$ by the distance formula and, when $T \neq O$, that $\overleftrightarrow{OT} \| \overleftrightarrow{PQ}$ as either both lines have no slope or else both have the same slope. Likewise, we have $\overleftrightarrow{OP} \| \overleftrightarrow{TQ}$. Since a quadrilateral is a parallelogram iff a pair of opposite sides are congruent and parallel, we can state part of our result above as follows.

Theorem 3.2. *Suppose A, B, C are noncollinear points. Then $\tau_{A,B} = \tau_{C,D}$ iff $\square CABD$ is a parallelogram.*

It follows that a translation moves each point the same distance in the same direction. For nonidentity translation $\tau_{A,B}$, the distance is given by AB and the direction by $\overrightarrow{AB}$. A visual image of a translation is suggested by Figure 3.2. You can probably see why translations are often called *slides* in elementary-school geometry.

We have yet to show that a translation is a collineation. Suppose line l has equation $aX + bY + c = 0$ and nonidentity translation $\tau_{P,Q}$ has

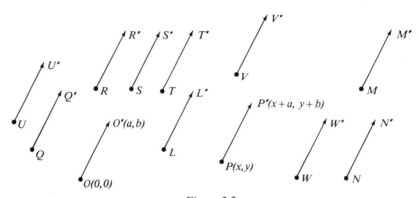

Figure 3.2

equations $x' = x + h$ and $y' = y + k$. So $\tau_{P,Q} = \tau_{O,R}$ where $R = (h, k)$ and $\overleftrightarrow{PQ} \| \overleftrightarrow{OR}$. Under the equations for $\tau_{P,Q}$, we see that $ax + by + c = 0$ iff $ax' + by' + (c - ah - bk) = 0$. We calculate that $\tau_{P,Q}(l)$ is the line m with equation $aX + bY + (c - ah - bk) = 0$. We have shown more than the fact that a translation is a collineation. By comparing the equations for lines l and m, we see l and m are parallel. Thus, a translation always sends a line to a parallel line. In general, a collineation α is a **dilatation** if $l \| \alpha(l)$ for every line l. Our calculation has shown that a translation is a dilatation. (While any collineation sends a *pair* of parallel lines to a pair of parallel lines, a dilatation sends *each* given line to a line parallel to the given line. For example, we shall see that a rotation of $90°$ is a collineation but not a dilatation.) We have even more results from our calculation. Now l and m are the same line iff $ah + bk = 0$. Since $\overleftrightarrow{OR}$ has equation $kX - hY = 0$, then $ah + bk = 0$ iff $l \| \overleftrightarrow{OR}$. Thus $\tau_{P,Q}(l) = l$ for line l iff $l \| \overleftrightarrow{PQ}$. We say transformation α **fixes** point P iff $\alpha(P) = P$. Transformation α **fixes** line l iff $\alpha(l) = l$ and, in general, **fixes** set s of points iff $\alpha(s) = s$. (Note that $\tau_{P,Q}$ fixes $\overleftrightarrow{PQ}$ but fixes no point on $\overleftrightarrow{PQ}$ when $P \neq Q$.) The next theorem summarizes the results of our calculation.

Theorem 3.3. *A translation is a dilatation. If $P \neq Q$, then $\tau_{P,Q}$ fixes no points and fixes exactly those lines that are parallel to $\overleftrightarrow{PQ}$.*

Dilatations are collineations. By the symmetry of parallelness for lines (i.e., $l \| l'$ implies $l' \| l$), the inverse of a dilatation is a dilatation. By the transitivity of parallelness for lines (i.e., $l \| l'$ and $l' \| l''$ implies $l \| l''$), the composite of two dilatations is a dilatation. So the dilatations form a group $\mathscr{D}$ of transformations, called **the dilatation group**.

We want to show the translations form a group $\mathscr{T}$ of transformations, called **the translation group**. Let $S = (a, b)$, $T = (c, d)$, and $R = (a + c, b + d)$. Then

$$\tau_{O,T}\tau_{O,S}((x, y)) = \tau_{O,T}((x + a, y + b)) = (x + a + c, y + b + d)$$
$$= \tau_{O,R}((x, y)).$$

From this we can draw several conclusions. Since $\tau_{O,T}\tau_{O,S} = \tau_{O,R}$, then a product of two translations is a translation. Also, by taking $R = O$, we see that the inverse of the translation $\tau_{O,S}$ with $S = (a, b)$ is $\tau_{O,T}$ with $T = (-a, -b)$. Hence, the set of all translations forms a group. Further, since $a + c = c + a$ and $b + d = d + b$, it follows that $\tau_{O,T}\tau_{O,S} = \tau_{O,S}\tau_{O,T}$. Therefore translations commute. Putting all this together, we have proved the following.

Theorem 3.4. *The translations form an abelian group $\mathcal{T}$. The dilatations form a group $\mathcal{D}$.*

The symbols "$\mathcal{T}$" and "$\mathcal{D}$" are reserved for the translation group $\mathcal{T}$ and the dilatation group $\mathcal{D}$ of Theorem 3.4.

§3.2 Halfturns

A *halfturn* turns out to be an involutory rotation, that is, a rotation of 180°. So a halfturn is just a special case of a rotation. Although we have not formally introduced rotations yet, we look at this special case now because halfturns are nicely related to translations and have such easy equations in the analytic geometry. We want to give a coordinate geometry definition for halfturns that is like the definition above for translations. Informally, we observe that if point A is rotated 180° about point P to point A', then P is the midpoint of A and A'. See Figure 3.3 below. Hence, we need only the midpoint formulas to obtain the desired equations. From equations $(x + x')/2 = a$ and $(y + y')/2 = b$ we can formulate our definition as follows. If $P = (a, b)$, then the **halfturn σ_P about point P** is the mapping with equations

$$\begin{cases} x' = -x + 2a, \\ y' = -y + 2b. \end{cases}$$

The use of the Greek letter sigma to denote halfturns will be explained later.

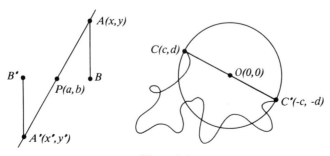

Figure 3.3

In particular, we note that for the halfturn about the origin we have $\sigma_O((x, y)) = (-x, -y)$. Under transformation σ_O does (x, y) go to $(-x, -y)$ by going directly through O, by rotating counterclockwise about O, by rotating clockwise about O, or by taking the other fanciful path illustrated in Figure 3.3? Either the answer is "None of the above" or perhaps it would be better to ask whether the question makes sense. Recall that transformations are just one-to-one correspondences among the points. There is actually no physical motion being described. (That is done in the study called *differential geometry*.) We might say we are describing the end position of physical motion. For example, the end position in rotating counterclockwise 180° and rotating clockwise 180° is the same even though these are physically different motions. Since our thinking is often aided by language indicating physical motion, we continue such usage as the customary "P goes to Q" in place of the more formal "P corresponds to Q."

What properties of a halfturn follow immediately from the definition of σ_P? First, for any point A, the midpoint of A and $\sigma_P(A)$ is P. This, of course, is how the definition was formulated in the first place. However, from this simple fact alone, it follows that σ_P is an involutory transformation. Also from this simple fact, it follows that σ_P fixes exactly the one point P. It even follows that σ_P fixes line l iff P is on l. We shall also see this ast result again in showing σ_P is a collineation. Suppose line l has equation $aX + bY + c = 0$. Let $P = (h, k)$. Then σ_P has equations $x' = -x + 2h$ and $y' = -y + 2k$. Then $ax + by + c = 0$ iff $ax' + by' + c - 2(ah + bk + c) = 0$. So $\sigma_P(l)$ is the line m with equation $aX + bY + c - 2(ah + bk + c) = 0$. Therefore, not only is σ_P a collineation but a dilatation as $l \| m$. Finally, l and m are the same line iff $ah + bk + c = 0$, which holds iff (h, k) is on l. We have proved the following.

Theorem 3.5. *A halfturn is an involutory dilatation. The midpoint of points A and $\sigma_P(A)$ is P. Halfturn σ_P fixes point A iff $A = P$. Halfturn σ_P fixes line l iff P is on l.*

Since a halfturn is an involution, then $\sigma_P\sigma_P = \iota$. What can be said about the product of two halfturns in general? Let $P = (a, b)$ and $Q = (c, d)$. Then

$$\sigma_Q\sigma_P((x, y)) = \sigma_Q((-x + 2a, -y + 2b))$$
$$= (-[-x + 2a] + 2c, -[-y + 2b] + 2d)$$
$$= (x + 2[c - a], y + 2[d - b]).$$

Since $\sigma_Q\sigma_P$ has equations $x' = x + 2(c - a)$ and $y' = y + 2(d - b)$, then $\sigma_Q\sigma_P$ is a translation. This proves the important result that the product of two halfturns is a translation. Suppose R is the point such that Q is the midpoint of P and R. Then

$$\sigma_Q\sigma_P(P) = \sigma_Q(P) = R \quad \text{and} \quad \sigma_R\sigma_Q(P) = \sigma_R(R) = R.$$

Since there is a unique translation taking P to R, then each of $\sigma_Q\sigma_P$ and $\sigma_R\sigma_Q$ must be $\tau_{P,R}$. We have proved the following.

Theorem 3.6. *If Q is the midpoint of points P and R, then*

$$\sigma_Q\sigma_P = \tau_{P,R} = \sigma_R\sigma_Q.$$

Observe that the theorem above states that a product of two halfturns is a translation and, conversely, a translation is a product of two halfturns. Also note that $\sigma_Q\sigma_P$ moves each point *twice* the directed distance from P to Q.

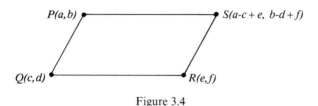

Figure 3.4

We now consider a product of three halfturns. By thinking about the equations, it should almost be obvious that $\sigma_R\sigma_Q\sigma_P$ is itself a halfturn. We shall prove that and a little more. Suppose $P = (a, b)$, $Q = (c, d)$, and $R = (e, f)$. Let $S = (a - c + e, b - d + f)$. In case, P, Q, R are not collinear, then $\square PQRS$ is a parallelogram. This is easy to check as opposite sides of the quadrilateral are congruent and parallel. See Figure 3.4. We calculated $\sigma_Q\sigma_P((x, y))$ above. Whether P, Q, R are collinear or not, with one more step we obtain

$$\sigma_R\sigma_Q\sigma_P((x, y)) = (-x + 2[a - c + e], -y + 2[b - d + f])$$
$$= \sigma_S((x, y)).$$

Our theorem now follows.

Theorem 3.7. *A product of three halfturns is a halfturn. In particular, if points P, Q, R are not collinear, then $\sigma_R\sigma_Q\sigma_P = \sigma_S$ where $\square PQRS$ is a parallelogram.*

We can solve the equation $\tau_{A,B} = \sigma_D\sigma_C$ for any one of A, B, C, D in terms of the other three. Knowing C, D, and one of A or B, we let the other be defined by the equation $\sigma_D\sigma_C(A) = B$ or the equivalent equation $\sigma_C\sigma_D(B) = A$. In either case, product $\sigma_D\sigma_C$ is the unique translation taking A to B, and so $\sigma_D\sigma_C = \tau_{A,B}$. When we know both A and B, we let M be the midpoint of A and B. So $\tau_{A,B} = \sigma_M\sigma_A$. Knowing A, B, D, we have C is the unique solution for Y in the equation $\sigma_D\sigma_M\sigma_A = \sigma_Y$ as then $\tau_{A,B} = \sigma_M\sigma_A = \sigma_D\sigma_Y$. Knowing A, B, C, we have D is the unique solution for Z in the equation $\sigma_M\sigma_A\sigma_C = \sigma_Z$ as then $\tau_{A,B} = \sigma_M\sigma_A = \sigma_Z\sigma_C$. We have proved the following theorem.

Theorem 3.8. *Given any three of the not necessarily distinct points A, B, C, D, then the fourth is uniquely determined by the equation* $\tau_{A,B} = \sigma_D\sigma_C$.

If $\sigma_Q\sigma_P = \tau_{P,R}$, then $\tau_{P,R}^{-1} = \sigma_P\sigma_Q$. So $\sigma_Q\sigma_P = \sigma_P\sigma_Q$ iff $P = Q$. However, a product of three halfturns can be written backwards since for any points P, Q, R there is a point S such that

$$\sigma_R\sigma_Q\sigma_P = \sigma_S = \sigma_S^{-1} = (\sigma_R\sigma_Q\sigma_P)^{-1} = \sigma_P^{-1}\sigma_Q^{-1}\sigma_R^{-1} = \sigma_P\sigma_Q\sigma_R.$$

Hence, although halfturns do not commute in general, we have proved the following theorem.

Theorem 3.9. $\sigma_R\sigma_Q\sigma_P = \sigma_P\sigma_Q\sigma_R$ *for any points P, Q, R.*

The halfturns do not form a group by themselves. A product of two halfturns is a translation. Since a translation is a product of two halfturns, then the product in either order of a translation and a halfturn is a halfturn by Theorem 3.7. In general, a product of an even number of halfturns is a product of translations and, hence, is a translation. Then, a product of an odd number of halfturns is a halfturn followed by a translation and, hence, is a halfturn. Thus the group generated by the halfturns contains just the halfturns and the translations.

Theorem 3.10. *The union of the translations and the halfturns forms a group $\mathcal{H}$.*

We reserve "$\mathcal{H}$" for the group in the theorem. This group seems to have no name other than ***the group generated by the halfturns.***

Can you define the following key words and symbols introduced in this chapter: *translation, halfturn, dilatation, α fixes s, $\tau_{P,Q}$, σ_R, $\mathcal{D}$, $\mathcal{H}$, $\mathcal{T}$?*

§3.3 Exercises

3.1. Find all triangles such that three given noncollinear points are the midpoints of the sides of the triangle.

3.2. Prove $\tau_{A,B}\sigma_P\tau_{A,B}^{-1} = \sigma_Q$ where $Q = \tau_{A,B}(P)$.

3.3. In Figure 3.5, sketch the shortest road from B to E that crosses the river r over a bridge at right angles to the parallel banks of the river r.

3.4. In Figure 3.5, sketch points X, Y, Z such that $\sigma_A\sigma_E\sigma_D = \sigma_X$, $\sigma_D\tau_{A,C} = \sigma_Y$, and $\tau_{B,C}\tau_{A,B}\tau_{E,A}(A) = Z$.

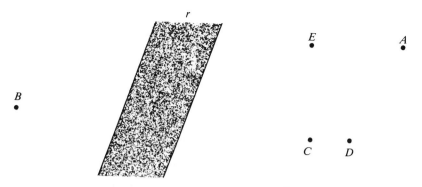

Figure 3.5

3.5. True or False
   (a) A product of two involutions is an involution or ι.
   (b) $\mathcal{D} \subset \mathcal{H} \subset \mathcal{T}$.
   (c) If δ is a dilatation and lines l and m are parallel, then $\delta(l)$ and $\delta(m)$ are parallel to l.
   (d) Given points A, B, C, there is a D such that $\tau_{A,B} = \tau_{D,C}$.
   (e) Given points A, B, C, there is a D such that $\tau_{A,B} = \sigma_D \sigma_C$.
   (f) If $\tau_{A,B}(C) = D$, then $\tau_{A,B} = \tau_{C,D}$.
   (g) If $\sigma_Q \sigma_P = \tau_{P,R}$, then $\sigma_P \sigma_Q = \tau_{R,P}$.
   (h) $\sigma_A \sigma_B \sigma_C = \sigma_B \sigma_C \sigma_A$ for points A, B, C.
   (i) A translation has equations $x' = x - a$ and $y' = y - b$.
   (j) $\sigma_Q \sigma_P = \tau_{P,Q}^2$ for any points P and Q.

3.6. $x' = -x + 3$ and $y' = -y - 8$ are the equations for which transformation? What are the equations for $\tau_{S,T}^{-1}$ if $S = (a, c)$ and $T = (g, h)$.

3.7. Prove or disprove: $\sigma_P \tau_{A,B} \sigma_P = \tau_{C,D}$ where $C = \sigma_P(A)$ and $D = \sigma_P(B)$.

3.8. If $P_i = (a_i, b_i)$ for $i = 1, 2, 3, 4, 5$, then what are the equations for the product

$$\tau_{P_4, P_5} \tau_{P_3, P_4} \tau_{P_2, P_3} \tau_{P_1, P_2} \tau_{O, P_1}?$$

3.9. What is the image of the line with equation $Y = 5X + 7$ under σ_P when $P = (-3, 2)$?

3.10. If α is a translation, show that $\alpha \sigma_P$ is the halfturn about the midpoint of points P and $\alpha(P)$. What is $\sigma_P \alpha$?

3.11. Use Theorem 3.9 twice to prove two translations commute.

3.12. Draw line l with equation $Y = 5X + 7$ and point P with coordinates $(2, 3)$. Then draw $\sigma_P(l)$.

3.13. Show $\tau_{P,Q}$ has infinite order if $P \neq Q$.

3.14. Suppose $\langle \tau_{P,Q} \rangle$ is a subgroup of $\langle \tau_{R,S} \rangle$. Show there is a positive integer n such that $PQ = nRS$.

3.15. Show $\langle \tau_{P,Q} \rangle = \langle \tau_{R,S} \rangle$ implies $\tau_{P,Q} = \tau_{R,S}$ or $\tau_{P,Q} = \tau_{S,R}$.

3.16. If $x' = ax + by + c$ and $y' = dx + ey + f$ are the equations for mapping α, then what are the necessary and sufficient conditions on the coefficients for α to be a transformation. Is such a transformation always a collineation?

3.17. Given $\angle ABC$, construct P on $\overleftrightarrow{AB}$ and Q on $\overleftrightarrow{BC}$ such that $PQ = AB$ and $\overleftrightarrow{PQ}$ intersects $\overleftrightarrow{BC}$ at an angle of $60°$.

3.18. Given two circles c_1 and c_2 and a segment $\overline{CD}$, construct points A on c_1 and B on c_2 such that $\overline{AB}$ is congruent and parallel to $\overline{CD}$.

Chapter 4

Reflections

§4.1 Equations for a Reflection

Suppose house H and barn B are on the same side of the Mississippi River m as indicated in Figure 4.1. The problem is to go from the house to the barn by way of the river along the shortest possible path. You may wish to stop and try to solve this problem before reading further. Supposedly, we are after a pail of water from the river. Anyway, it is clear that we do not walk along the river once we get there. Suppose the shortest path touches the river at point R. So R is the point on m such that $HR + RB$ is minimal. As you may have guessed, a solution to this problem has something to do with reflections. If we let H' be the mirror image of H reflected in line m, then surely $HR = H'R$. So we want R on m such that $H'R + RB$ is minimal. This problem is very easy. Obviously R is the intersection of m and $\overline{BH'}$. The path obtained is the path of a ray of light traveling from H to B that is reflected in mirror m, as the angle of incidence at m is congruent to the angle of reflection.

It is almost certain that you can point to H' in the problem above. Points H and H' are symmetrically placed with respect to line m. However, it may not be so easy to describe exactly where H' is located in a succinct manner.

H •

• B

m

Figure 4.1

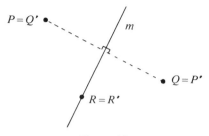

Figure 4.2

In Figure 4.2, point Q is the reflected image of point P in line m. Check to see if the following definition satisfies your intuitive idea of just what a reflection should be. **Reflection σ_m in line** m is the mapping defined by

$$\sigma_m(P) = \begin{cases} P, & \text{if point } P \text{ is on } m, \\ Q, & \text{if } P \text{ is off } m \text{ and } m \text{ is the} \\ & \text{perpendicular bisector of } \overline{PQ}. \end{cases}$$

Some properties of a reflection follow immediately from the definition. First, $\sigma_m \neq \iota$ but $\sigma_m^2 = \iota$ as the perpendicular bisector of $\overline{PQ}$ is the perpendicular bisector of $\overline{QP}$. Hence, σ_m is onto as $\sigma_m(P)$ is the point mapped onto given point P since $\sigma_m(\sigma_m(P)) = P$ for any point P. Also, σ_m is one-to-one as $\sigma_m(A) = \sigma_m(B)$ implies $\sigma_m(\sigma_m(A)) = \sigma_m(\sigma_m(B))$ and $A = B$. Therefore σ_m is an involutory transformation. Then, from the definition of σ_m, it follows that σ_m interchanges the halfplanes of m. Clearly σ_m fixes point P iff P is on m. Not only does σ_m fix line m, but σ_m fixes every point on m. In general, transformation α is said to fix **pointwise** set s of points if $\alpha(P) = P$ for every point P in s. Note the difference between fixing a set and fixing a set pointwise. Every line perpendicular to m is fixed by σ_m, but none of these lines is fixed pointwise as each contains only one fixed point. Suppose line l is distinct from m and is fixed by σ_m. Let $Q = \sigma_m(P)$ for some point P that is on l but off m. Then P and Q are both on l since l is fixed, and m is the perpendicular bisector of $\overline{PQ}$. Hence, l and m are perpendicular. These immediate properties of a reflection are summarized in the following theorem.

Theorem 4.1. *Reflection σ_m is an involutory transformation that interchanges the halfplanes of m. Reflection σ_m fixes point P iff P is on m. Reflection σ_m fixes line l pointwise iff $l = m$. Reflection σ_m fixes line l iff $l = m$ or $l \perp m$.*

We do not use the word *reflection* to denote the image of a point or of a set of points. A reflection is a transformation and never a set of points. Point $\sigma_m(P)$ is the *image* of point P under the *reflection σ_m*.

You will have noticed that we have used the Greek letter sigma, σ, for both halfturns and reflections. This follows international custom and is a convenient convention for English as then rho, ρ, is left free for use later with rotations. The Greek σ corresponds to the Roman s which begins the

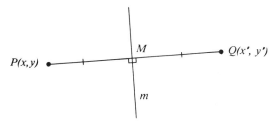

Figure 4.3

German word *Spiegelung*, meaning *reflection*. A halfturn is a sort of "reflection in a point," although we shall never use this terminology and thus avoid confusion. The similar notation for halfturns and reflections emphasizes the important property they do share, namely that of being involutions:

$$\sigma_l = \sigma_l^{-1}, \qquad \sigma_P = \sigma_P^{-1}.$$

What are the equations for a reflection? We shall find the equations for σ_m where m has equation $aX + bY + c = 0$. Let $P = (x, y)$ and $\sigma_m(P) = (x', y') = Q$. See Figure 4.3. For the moment, suppose P is off m. Now, the line through points (x, y) and (x', y') is perpendicular to line m. This geometric fact is expressed algebraically by the equation

$$b(x' - x) = a(y' - y).$$

Also, $((x + x')/2, (y + y')/2)$ is the midpoint of $\overline{PQ}$ and is on line m. This geometric fact is expressed algebraically by the equation

$$a\left(\frac{x + x'}{2}\right) + b\left(\frac{y + y'}{2}\right) + c = 0.$$

Rewriting these two equations as

$$\begin{cases} bx' - ay' = bx - ay, \\ ax' + by' = -2c - ax - by, \end{cases}$$

we see we have two linear equations in the two unknowns x' and y'. Solving these equations for x' and y', we get

$$x' = \frac{b(bx - ay) + a(-2c - ax - by)}{b^2 + a^2}$$

$$= \frac{b^2x + a^2x - 2a^2x - 2aby - 2ac}{a^2 + b^2},$$

$$y' = \frac{b(-2c - ax - by) - a(bx - ay)}{b^2 + a^2}$$

$$= \frac{a^2y + b^2y - 2b^2y - 2bax - 2bc}{a^2 + b^2}.$$

With these equations in the form

$$\begin{cases} x' = x - \dfrac{2a(ax + by + c)}{a^2 + b^2}, \\ y' = y - \dfrac{2b(ax + by + c)}{a^2 + b^2}, \end{cases}$$

it is easy to check that the equations also hold when P is on m as then $ax + by + c = 0$. This proves the theorem.

Theorem 4.2. *If line m has equation $aX + bY + c = 0$, then σ_m has equations*

$$\begin{cases} x' = x - 2a(ax + by + c)/(a^2 + b^2), \\ y' = y - 2b(ax + by + c)/(a^2 + b^2). \end{cases}$$

You can now appreciate our synthetic definition of *reflection*. Suppose we had defined a reflection as a transformation having equations given by Theorem 4.2. Not only would this have seemed artificial, since these equations are not something you would think of examining in the first place, but just imagine trying to prove Theorem 4.1 from these equations. To show σ_m is an involution, we would need to show $(x')' = x$ and $(y')' = y$. Although this is conceptually easy, the actual computation involves a considerable amount of algebra.

§4.2 Properties of a Reflection

We have already mentioned those properties of a reflection that follow immediately from the definition (Theorem 4.1). Another important property is that a reflection *preserves distance*, which means the distance from $\sigma_m(P)$ to $\sigma_m(Q)$ is equal to the distance from P to Q for all points P and Q. The name for any transformation that preserves distance comes from the Greek *isos* (equal) and *metron* (measure). Thus, in general, a transformation α is an *isometry* if $P'Q' = PQ$ for all points P and Q where $P' = \alpha(P)$ and $Q' = \alpha(Q)$. We want to prove σ_m is an isometry. One method is simply to use the equations for a reflection (Theorem 4.2) and show that the distance from (x'_1, y'_1) to (x'_2, y'_2) is equal to the distance from (x_1, y_1) to (x_2, y_2). Although this entails only algebra of the most elementary kind, there is so much of it that this method is rejected.

To show σ_m is an isometry for any line m, we shall consider several cases. Suppose P and Q are two points, $P' = \sigma_m(P)$, and $Q' = \sigma_m(Q)$. We must show $P'Q' = PQ$. If $\overleftrightarrow{PQ} = m$ or if $\overleftrightarrow{PQ} \perp m$, then the desired result follows immediately from the definition of σ_m. Also, if $\overleftrightarrow{PQ}$ is parallel to m but distinct from m, the result follows easily as $\square PQQ'P'$ is a rectangle and so opposite sides $\overline{PQ}$ and $\overline{P'Q'}$ are congruent. Further, if one of P or Q, say P,

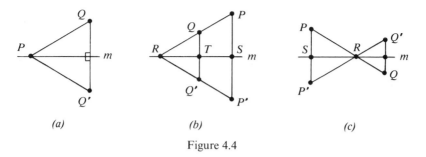

Figure 4.4

is on m and Q is off m, then $P'Q' = PQ$ follows from the fact $P' = P$ and that m is the locus of all points equidistant from Q and Q'. See Figure 4.4a. Finally, suppose P and Q are both off m and that $\overleftrightarrow{PQ}$ intersects m at point R but is not perpendicular to m. So $RP = RP'$ and $RQ = RQ'$. The desired result, $P'Q' = PQ$, then follows provided R, P', Q' are shown to be collinear. See Figure 4.4b for the case P and Q are on the same side of m and Figure 4.4c for the case P and Q are on opposite sides of m. Let S be the midpoint of P and P'; let T be the midpoint of Q and Q'. So $m = \overleftrightarrow{ST}$. In either case, each of the angles $\angle SRP'$, $\angle SRP$, $\angle TRQ$, and $\angle TRQ'$, is congruent to the next. That $\angle SRP'$ is congruent to $\angle TRQ'$ implies R, P', Q' are collinear. So $P'Q' = PQ$, as desired.

Theorem 4.3. *Reflection σ_m is an isometry.*

Now that we know a reflection is an isometry, a long sequence of other properties dependent only on distance will follow. Since these properties are shared by all isometries, we shall consider a general isometry α. Suppose A, B, C are any three points and let $A' = \alpha(A)$, $B' = \alpha(B)$, $C' = \alpha(C)$. Since α preserves distance, if $AB + BC = AC$ then $A'B' + B'C' = A'C'$ as $A'B' = AB$, $B'C' = BC$, and $A'C' = AC$. Hence, A–B–C implies A'–B'–C'; in other words, if B is between A and C, then B' is between A' and C'. We describe this by saying α *preserves betweenness*. The special case $AB = BC$ in the argument above implies $A'B' = B'C'$. In other words, if B is the midpoint of A and C, then B' is the midpoint of A' and C'. Thus we say α *preserves midpoints*. More generally, since $\overline{AB}$ is the union of A, B, and all points between A and B, then $\alpha(\overline{AB})$ is the union of A', B', and all points between A' and B'. So $\alpha(\overline{AB}) = \overline{A'B'}$ and we say α *preserves segments*. Likewise, since α is onto by definition and $\overrightarrow{AB}$ is the union of $\overline{AB}$ and all points C such that A–B–C, then $\alpha(\overrightarrow{AB})$ is the union of $\overline{A'B'}$ and all points C' such that A'–B'–C'. So $\alpha(\overrightarrow{AB}) = \overrightarrow{A'B'}$ and we say α *preserves rays*. Since $\overleftrightarrow{AB}$ is the union of $\overrightarrow{AB}$ and $\overrightarrow{BA}$, then $\alpha(\overleftrightarrow{AB})$ is the union of $\overrightarrow{A'B'}$ and $\overrightarrow{B'A'}$, which is $\overleftrightarrow{A'B'}$. So α is a transformation that preserves lines; in other words, α is a collineation. If A, B, C are not collinear, then $AB + BC > AC$ and so $A'B' + B'C' > A'C'$ and A', B', C' are not collinear. Then, since $\triangle ABC$ is a union of the three segments $\overline{AB}, \overline{BC}, \overline{CA}$, then we conclude $\alpha(\triangle ABC)$

is just $\triangle A'B'C'$. So an isometry *preserves triangles*. It follows that α *preserves angles* as $\alpha(\angle ABC) = \angle A'B'C'$. Not only does α preserve angles but α also *preserves angle measure*. That is, $m\angle ABC = m\angle A'B'C'$ since $\triangle ABC \cong \triangle A'B'C'$ by *SSS*. Finally, if $\overrightarrow{BA} \perp \overrightarrow{BC}$ then $\overrightarrow{B'A'} \perp \overrightarrow{B'C'}$ since $m\angle ABC = 90$ implies $m\angle A'B'C' = 90$. So α *preserves perpendicularity*. Also, α preserves parallelness, since parallel lines have a common perpendicular, but we already know α preserves parallelness since α is a collineation. All this is put together as our new theorem.

Theorem 4.4. *An isometry is a collineation that preserves betweenness, midpoints, segments, rays, triangles, angles, angle measure, and perpendicularity.*

Isometries do all that! Isometries are very important, and we shall continue to study them for some time.

Perhaps you remember making ink blots as a child. It's rather fun and you might try it again. The idea is to drop some ink on a sheet of paper, fold the paper with the ink on the inside, press the sheet firmly, and unfold to reveal the ink blot. The fold is a *line of symmetry* for the ink blot. We formalize and extend this idea as follows. Line *m* is a **line of symmetry** for set *s* of points if $\sigma_m(s) = s$, that is, if σ_m fixes *s*. Point *P* is a **point of symmetry** for set *s* of points if $\sigma_P(s) = s$. Isometry α is a **symmetry** for set *s* of points if $\alpha(s) = s$. In Figure 4.5, line *m* is one of six lines of symmetry for the regular hexagon and point *P* is a point of symmetry for the regular hexagon. It is clear that a rotation of 60° about *P* is also a symmetry for the regular hexagon. Note that the shaded hexagonal region in the figure has no line of symmetry and no point of symmetry, although the shaded region does have the symmetry of a rotation of 120°. The center of this rotation will be defined later as a *center of symmetry*. This is mentioned now only to prevent forming a misconception of the definition of a point of symmetry.

Let's consider the symmetries of the rectangle that is not a square in Figure 4.6. Evidently the axes of the plane are lines of symmetry for the rectangle and the origin is a point of symmetry for the rectangle. Denoting

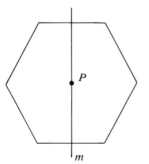
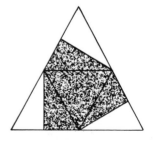

Figure 4.5

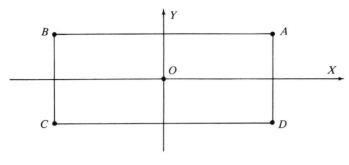

Figure 4.6

the reflection in the X-axis by σ_h and the reflection in the Y-axis by σ_v, we then have that σ_h, σ_v, σ_O, and $\imath$ are symmetries for the rectangle. Note that $\imath$ is a symmetry for any set of points. Since the image of the rectangle is known once it is known which of A, B, C, D is the image of A, then these four are the only possible symmetries for the rectangle. Recall that these isometries are the elements of group V_4 introduced in Chapter 2. See Table 2.2 in Chapter 2. We see that the symmetries for a rectangle that is not a square form a group. The fact that the symmetries of any set of points form a group is shown next.

Let s be any set of points. The set of symmetries for s is not empty as $\imath$ is a symmetry for s. Suppose α and β are symmetries for s. Then $\beta\alpha(s) = \beta(\alpha(s)) = \beta(s) = s$. So the set of symmetries has the closure property. If α is a symmetry for s, then α and α^{-1} are transformations and $\alpha^{-1}(s) = \alpha^{-1}(\alpha(s)) = \imath(s) = s$. So the set of symmetries also has the inverse property. We have proved the following.

Theorem 4.5. *The set of all symmetries of a set of points forms a group.*

What happens if the set of points in this theorem is taken to be the set of all points? In this special case, the symmetries are exactly the same thing as the isometries. The theorem thus provides the following corollary.

Theorem 4.6. *The set of all isometries forms a group.*

The group of all symmetries for a set s of points is called the **symmetry group** of s or the **full group of symmetries** for s. The group of all isometries is denoted by $\mathscr{I}$. So $\mathscr{I}$ is the symmetry group of the plane, and V_4 is the symmetry group for a rectangle that is not a square. In case you are wondering why "V" is always used to denote this symmetry group, the group is also known as Klein's *Vierergruppe* (*four-group* in German). Felix Klein (1849–1925) was interested in applying the concept of a group as a convenient means of characterizing the various geometries of his time. His models of non-Euclidean geometries eventually helped mathematicians understand the significance and place of the new geometries within mathematics. Upon accepting a chair at the University of Erlangen, in 1872 Klein made his famous inaugural address, which has become known as the *Erlanger Program*. The Erlanger Program describes a geometry by its group of symmetries. For example, basic Euclidean geometry is characterized by its isometries, those transformations that preserve distance and, hence, are concerned with such properties as length, congruence, angle measure, and collinearity. Much of our work is consistent with this approach. Klein also played an important role in the development of mathematics in the United States through his many American students.

Can you define the following key words and symbols introduced in this chapter: *reflection, isometry, symmetry, point of symmetry, line of symmetry, symmetry group, fixed pointwise, σ_m, $\mathcal{I}$*?

§4.3 Exercises

4.1. Prove: If $\alpha \in \mathcal{H}$, then α is an isometry.

4.2. What is the minimum length of a flat-against-the-wall, full-length mirror for the Smiths who range in height from 170 cm to 182 cm, if you assume eyes are 10 cm below the top of the head?

4.3. Fill in the missing entry in each row:

Equation of line m	Point P	Point $\sigma_m(P)$
$X = 0$	(x, y)	*
$Y = 0$	*	(x, y)
$Y = X$	*	$(2, 3)$
$Y = X$	(x, y)	*
$X = 2$	$(-2, 3)$	*
$Y = -3$	$(-4, -1)$	*
$Y = -3$	(x, y)	*
*	$(5, 3)$	$(-8, 3)$
*	$(0, 3)$	$(-3, 0)$
*	$(-y, -x)$	(x, y)
$Y = 2X$	*	$(4, 3)$

4.4. What is the symmetry group of a parallelogram that is neither a rectangle nor a rhombus? What is the symmetry group of a rhombus that is not a square?

4.5. Given two parallel lines p and q in Figure 4.7, sketch a construction of the path of a ray of light issuing from A and which passes through E after being reflected exactly twice in p and exactly once in q.

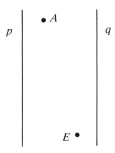

Figure 4.7

4.6. True or False
(a) If isometry α interchanges distinct points P and Q, then α fixes the midpoint of P and Q.
(b) If P is a point of symmetry for set s of points, then P is in s.
(c) If l and m are perpendicular lines, then σ_l is a line of symmetry for m.
(d) $\sigma_p = \sigma_P^{-1}$ if point P is on line p.
(e) If $\alpha \in \mathscr{H}$ but $\alpha \neq \iota$, then $\langle \alpha \rangle$ has order 2.
(f) Reflection σ_m fixes the halfplanes of m but does not fix the halfplanes pointwise.
(g) Reflection σ_m fixes line l iff $l \perp m$.
(h) For line l and point P, $\sigma_l = \sigma_l^{-1} \neq \iota$ and $\sigma_P = \sigma_P^{-1} \neq \iota$.
(i) A regular pentagon has a point of symmetry.
(j) The symmetry group of a rectangle has order 4.

4.7. What are the images of $(0, 0)$, $(1, -3)$, $(-2, 1)$, and $(2, 4)$ under the reflection in the line with equation $Y = 2X - 5$?

4.8. Why can't a letter of the alphabet have two points of symmetry?

4.9. What are the symmetries for the various playing cards in a standard bridge deck? What are the symmetries if the indices are ignored?

4.10. What capital letters could be cut out of paper and given a single fold to produce Figure 4.8?

Figure 4.8

4.11. A cup of coffee and a cup of milk contain equal amounts of liquid. A spoonful of milk is transferred from the milk cup to the coffee cup, and the coffee cup is stirred. Then, a spoonful of the mixture is returned to the milk cup. The two cups again have the same amount of liquid. Now, is there more milk in the coffee cup or more coffee in the milk cup?

4.12. If mapping α is such that $A'B' = AB$ for all points A and B where $A' = \alpha(A)$ and $B' = \alpha(B)$, then show α is an isometry.

4.13. Given line m and three points P, Q, $\sigma_m(P)$, then construct with a ruler only the point $\sigma_m(Q)$.

4.14. Given a line and two circles, construct the squares having a pair of opposite vertices on the line and a vertex on each of the two circles.

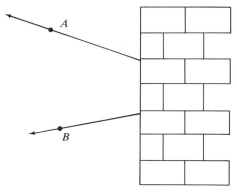

Figure 4.9

4.15. The vertex of $\angle AVB$ is obstructed in Figure 4.9. Without using the region behind the obstruction, construct that part of the angle bisector that is not behind the obstruction.

Chapter 5

Congruence

§5.1 Isometries as Products of Reflections

The halfturns generate the group $\mathcal{H}$. What group of isometries does the set of reflections generate? Since a reflection is its own inverse, every element in this group must be a product of reflections (Theorem 2.4). A product of reflections is clearly an isometry; in this section we show that every isometry is a product of reflections. Thus, we shall see that the reflections generate all of $\mathcal{I}$. The reflections are the building blocks for the symmetries of the plane.

Suppose you were asked to actually prove the obvious fact that σ_m and σ_c are not equal. You could answer that σ_m and σ_c must be distinct since they have different sets of fixed points. Looking at the fixed points of isometries turns out to be very rewarding in general. Specifically, this attack is used below to show the reflections generate all the isometries.

Knowing point P is on the line through distinct points A and B and knowing the nonzero distance AP, we do not know which of two possible points is P. However, if we also know the distance BP, then P is uniquely determined. It follows that an isometry fixing both A and B must also fix the point P, since an isometry is a collineation that preserves distance. In other words, an isometry fixing distinct points A and B must fix every point on the line through A and B. Suppose now that an isometry fixes each of three noncollinear points A, B, C. Then we have just observed the isometry must fix every point on $\triangle ABC$ as the isometry fixes every point on any one of the lines $\overleftrightarrow{AB}$, $\overleftrightarrow{BC}$, $\overleftrightarrow{CA}$. Every point Q in the plane lies on a line that intersects $\triangle ABC$ in two distinct points; for example, a line through Q and the midpoint M of A and B must intersect the triangle at another point of the triangle different from M. Hence the point Q is on a line containing two fixed points and, therefore, must also be fixed. So an isometry that fixes three noncollinear points must fix every point Q in the plane. We state these results as our first theorem.

Theorem 5.1. *If an isometry fixes two distinct points on a line, then the isometry fixes that line pointwise. If an isometry fixes three noncollinear points, then the isometry must be the identity.*

Suppose isometries α and β are such that $\alpha(P) = \beta(P)$, $\alpha(Q) = \beta(Q)$, and $\alpha(R) = \beta(R)$ for three noncollinear points P, Q, R. Multiplying each of these equations above by β^{-1} on the left, we see that $\beta^{-1}\alpha$ fixes each of the three noncollinear points P, Q, R. Then $\beta^{-1}\alpha = \iota$ by the first theorem. Multiplying this last equation by β on the left, we have $\alpha = \beta$. We have proved a powerful theorem.

Theorem 5.2. *If α and β are isometries such that*

$$\alpha(P) = \beta(P), \qquad \alpha(Q) = \beta(Q), \quad and \quad \alpha(R) = \beta(R)$$

for three noncollinear points P, Q, R, then $\alpha = \beta$.

Suppose isometry α fixes distinct points P and Q on line m. We know two possibilities for α, namely ι and σ_m. We shall show these are the only two possibilities by supposing $\alpha \neq \iota$ and proving $\alpha = \sigma_m$. If $\alpha \neq \iota$, then there is a point R not fixed by α. So R is off m (Theorem 5.1), and P, Q, R are three noncollinear points. Let $R' = \alpha(R)$. So $PR = PR'$ and $QR = QR'$, as α is an isometry. Therefore, m is the perpendicular bisector of $\overline{RR'}$ as each of P and Q is in the locus of all points equidistant from R and R'. Hence, $\alpha(R) = R' = \sigma_m(R)$ as well as $\alpha(P) = P = \sigma_m(P)$ and $\alpha(Q) = Q = \sigma_m(Q)$. By the previous theorem, we have $\alpha = \sigma_m$ and the following theorem.

Theorem 5.3. *An isometry that fixes two points is a reflection or the identity.*

Suppose isometry α fixes exactly one point C. Let P be a point different from C, let $\alpha(P) = P'$, and let m be the perpendicular bisector of $\overline{PP'}$. Since $CP = CP'$ as α is an isometry, then C is on m. So $\sigma_m(C) = C$ and $\sigma_m(P') = P$. Then $\sigma_m\alpha(C) = \sigma_m(C) = C$ and $\sigma_m\alpha(P) = \sigma_m(P') = P$. By the previous theorem, $\sigma_m\alpha = \iota$ or $\sigma_m\alpha = \sigma_l$ where $l = \overleftrightarrow{CP}$. However, $\sigma_m\alpha \neq \iota$ as otherwise α is σ_m and fixes more points than C. Thus $\sigma_m\alpha = \sigma_l$ for some line l. Multiplying this equation by σ_m on the left, we have $\alpha = \sigma_m\sigma_l$ and the next theorem.

Theorem 5.4. *An isometry that fixes exactly one point is a product of two reflections.*

Since $\iota = \sigma_m\sigma_m$ for any line m, we have the following theorem as a corollary of the previous two theorems.

Theorem 5.5. *An isometry that fixes a point is a product of at most two reflections.*

We are now prepared to prove the principal proposition that every isometry is a product of reflections. Actually, we shall do better than that by showing the product has at most three factors. (We count the number of factors even though the factors themselves may not be distinct.) The identity is a product of two reflections. Suppose nonidentity isometry α sends point P to different point Q. Let m be the perpendicular bisector of $\overline{PQ}$. Then $\sigma_m \alpha$ fixes point P. We have just seen in the theorem above that $\sigma_m \alpha$ must be a product β of at most two reflections. Hence $\alpha = \sigma_m \beta$ and α is a product of at most three reflections. Not only is every isometry α a product of reflections, but every isometry is a product of reflections where the number of factors in the product is 1, 2, or 3. Our initial observation in this section was that a product of reflections is an isometry. We have just shown the important converse of this observation.

Theorem 5.6. *A product of reflections is an isometry. Every isometry is a product of at most three reflections.*

From the theorem it follows that a given product of eight reflections is equal to a product of at most three reflections. The proof of our next theorem shows how you might find these one, two, or three reflections if you know the images of three noncollinear points P, Q, R under the given product of eight reflections. Suppose $\triangle PQR \cong \triangle ABC$. We know there is at most one isometry α such that $\alpha(P) = A$, $\alpha(Q) = B$, and $\alpha(R) = C$ (Theorem 5.2). The question is whether there exists at least one such isometry α. To see how we are going to show there is such an isometry, look at Figure 5.1. The reflection in l will take $\triangle PQR$ to $\triangle AQ_1R_1$, the reflection in m will take $\triangle AQ_1R_1$ to $\triangle ABR_2$, and the reflection in n will take $\triangle ABR_2$ to $\triangle ABC$. The composite will then take $\triangle PQR$ to $\triangle ABC$ in the desired manner. If certain points coincide, we may not need all three reflections. Used several times in the proof are the fact that a reflection is an isometry and the

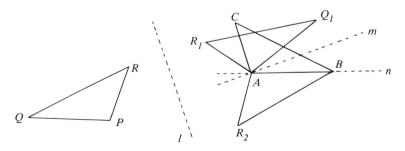

Figure 5.1

fact that the locus of all points equidistant from two points M and N is the perpendicular bisector of $\overline{MN}$.

Suppose $\triangle PQR \cong \triangle ABC$. So $AB = PQ$, $AC = PR$, and $BC = QR$. If $P \neq A$, then let $\alpha_1 = \sigma_l$ where l is the perpendicular bisector of $\overline{PA}$. If $P = A$, then let $\alpha_1 = \iota$. In either case, then $\alpha_1(P) = A$. Let $\alpha_1(Q) = Q_1$ and $\alpha_1(R) = R_1$. If $Q_1 \neq B$, then let $\alpha_2 = \sigma_m$ where m is the perpendicular bisector of $\overline{Q_1 B}$. In this case, point A is on m as $AB = PQ = AQ_1$. If $Q_1 = B$, then let $\alpha_2 = \iota$. In either case, we have $\alpha_2(A) = A$ and $\alpha_2(Q_1) = B$. Let $\alpha_2(R_1) = R_2$. If $R_2 \neq C$, then let $\alpha_3 = \sigma_n$ where n is the perpendicular bisector of $\overline{R_2 C}$. In this case, $n = \overleftrightarrow{AB}$ as $AC = PR = AR_1 = AR_2$ and $BC = QR = Q_1 R_1 = BR_2$. If $R_2 = C$, then let $\alpha_3 = \iota$. In any case, we have $\alpha_3(A) = A$, $\alpha_3(B) = B$, and $\alpha_3(R_2) = C$. Let $\alpha = \alpha_3 \alpha_2 \alpha_1$. Then

$$\alpha(P) = \alpha_3 \alpha_2 \alpha_1(P) = \alpha_3 \alpha_2(A) = \alpha_3(A) = A,$$

$$\alpha(Q) = \alpha_3 \alpha_2 \alpha_1(Q) = \alpha_3 \alpha_2(Q_1) = \alpha_3(B) = B,$$

$$\alpha(R) = \alpha_3 \alpha_2 \alpha_1(R) = \alpha_3 \alpha_2(R_1) = \alpha_3(R_2) = C,$$

as desired.

Theorem 5.7. *If* $\triangle PQR \cong \triangle ABC$, *then there is a unique isometry* α *such that* $\alpha(P) = A$, $\alpha(Q) = B$, *and* $\alpha(R) = C$.

Since two congruent segments are corresponding sides of congruent equilateral triangles and since two congruent angles are corresponding angles of congruent triangles, the theorem above has the theorem below as an immediate corollary.

Theorem 5.8. *Two segments, two angles, or two triangles are, respectively, congruent iff there is an isometry taking one to the other.*

In elementary geometry there are three different relations indicated by the same words *is congruent to*, one for segments, one for angles, and a third for triangles. All three can be combined under a generalized definition that applies to arbitrary sets of points as follows. If s_1 and s_2 are sets of points, then s_1 and s_2 are said to be **congruent** if there is an isometry α such that $\alpha(s_1) = s_2$.

§5.2 Paper Folding Experiments and Rotations

Paper folding experiments lead to conjectures in transformation geometry. Here we describe the experiments that lead to the conjectures; in the next chapter proofs are given to show the conjectures are actually theorems.

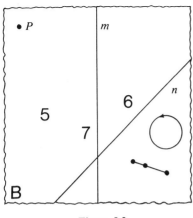

Figure 5.2

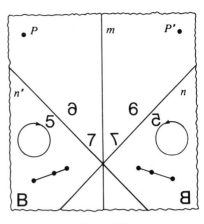

Figure 5.3

The only materials needed are a pencil and at least three sheets of waxed paper about 30 cm square. A ruler and a protractor might also help.

The first sheet of waxed paper is used to introduce the technique of using waxed paper to illustrate reflections. Take a sheet of waxed paper and fold the sheet in half over itself and make a crease. You might consider how many ways this can be done if you start with a perfect square. Not that it makes much difference, we suppose you have formed regions that are approximately rectangular rather than triangular. When you unfold the sheet, the crease of the fold is still evident. The crease represents a line which we shall call *m*. With the pencil, mark any point *P* off line *m*. Is it obvious how to find the mirror image *P'* of point *P* under the reflection in line *m*? We merely have to fold the sheet on *m*, trace the point *P* with the pencil (from either side of the sheet), unfold the sheet, and label the new point *P'*. To practice finding the images of figures under the reflection in line *m*, give yourself some figures such as those in Figure 5.2. After performing the operations described above, you should have Figure 5.3. You can discern many of the properties of a reflection by considering these figures.

For more practice, now trace the images of the images of the original numerals 5, 6, and 7 under the reflection in the line *n* shown in Figure 5.3. This will probably convince you that this tracing is not as easy as it sounds at first. You should not be impatient with yourself when you find you have traced a wrong figure. You may want to use an additional sheet to practice finding the image of one figure at a time under reflection first in line *m* and then the image of this image under reflection in line *n*.

What is the result of successive reflections in more than one line? This question is the subject of our investigation. We start by considering two lines. Of course, there are two cases: the lines may be parallel or the lines may intersect. We begin with the case of two parallel lines and a fresh sheet of waxed paper. However, before we can get underway, we are faced with the problem of constructing the two parallel lines. Using the theorem that two

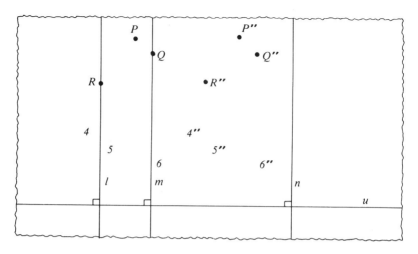

Figure 5.4

lines perpendicular to a third line are parallel, we can construct the parallel lines as follows. Fold a line u across the bottom of the second sheet of waxed paper. (If an edge of the sheet is straight, then this edge may be taken as u.) By folding the paper so that line u falls on itself, the line of the crease is perpendicular to line u. With two such foldings we have our parallel lines.

In order to make a fairly uncomplicated figure, we suppose you have folded your second sheet of waxed paper so that parallel lines l, m, and n are about 7, 11, and 21 cm, respectively, from the left-hand edge of the sheet. With the pencil, add some points such as P, Q, and R in Figure 5.4 and also some figures such as the numerals 4, 5, and 6, placed about 6, 8 and 12 cm from the left edge. These are all on the left-hand side of Figure 5.4. Now, by paper folding, find the images of these three points and three figures under the reflection in line l. These images are not given in Figure 5.4. Then find the images of these images under the reflection in line m. These images of the images are shown in Figure 5.4. Now ask yourself, "What do I conjecture is the result of first reflecting in line l and then reflecting in line m?" Although the "proper" conjecture is more evident in Figure 5.4 because the intermediate images are not shown, it probably does not take too long to see that the figures have just been "slid over." In what direction? And, how far? Study the other side of the sheet too. You should conjecture $\sigma_m \sigma_l$ is the translation through twice the directed distance from l to m.

That's the easy part. Now, reflect the last set of images in the line n. Ask yourself, "What is the result of composing the three reflections in the three parallel lines l, m, and n?" The answer will probably not come as quickly as the previous conjecture. After some study, the principal conjecture to be made here can be strengthened by one more folding of the waxed paper. As another hint: if the conjecture is correct, we need not go on with the experiment to consider reflections in four or more parallel lines.

For the last experiment, we want to make conjectures about composing reflections in lines that are concurrent. Take the third sheet of waxed paper and fold any three concurrent lines l, m, and n. Add some points, say one on l, one on m, and at least one off both l and m. You might also add some additional figures such as numerals or your initials. Now, by paper folding, find the images of these figures under the reflection in l. Then find the images of these images under the reflection in line m. It's time to stop and make some conjectures about the nature of the composition of two reflections in intersecting lines. Here is where the protractor comes in handy. Once you have done this, you are ready to reflect the last set of images in the line n. A conjecture about the composite of three reflections in three concurrent lines finishes the experiments. Again, you can strengthen your last conjecture by folding the sheet one more time.

Of course, you may not use any of your conjectures until they are proved as theorems. However, one of your conjectures should have involved *rotations*, which we now formally define in the most elementary manner. A *rotation* about point C through directed angle of $\Theta°$ is the transformation $\rho_{C,\Theta}$ that fixes C and otherwise sends a point P to the point P' where $CP' = CP$ and Θ is the directed angle measure of the directed angle from $\overrightarrow{CP}$ to $\overrightarrow{CP'}$. We agree that $\rho_{C,0}$ is the identity ι. Rotation $\rho_{C,\Theta}$ is said to have *center* C and *directed angle* $\Theta°$. We want to show a rotation is an isometry. Suppose $\rho_{C,\Theta}$ sends points P and Q to points P' and Q', respectively. If C, P, Q are collinear, then $PQ = P'Q'$ by the definition. If C, P, Q are not collinear, then $\triangle PCQ \cong \triangle P'C'Q'$ by *SAS* and $PQ = P'Q'$. See Figure 5.5. So $\rho_{C,\Theta}$ is a transformation that preserves distance.

Theorem 5.9. *A rotation is an isometry.*

For distinct points C and P, circle C_P is defined to be the circle with center C and radius CP. So $\overline{CP}$ is a radius of the circle C_P, and point P is on the circle.

That $\rho_{C,180} = \sigma_C$ follows from the fact that each of the transformations fixes point C and otherwise sends any point P to a point P' such that C is the

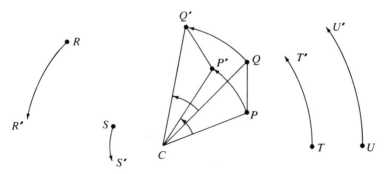

Figure 5.5

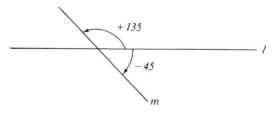

Figure 5.6

midpoint of P and P'. The remainder of the following theorem also follows immediately from the definition of a rotation.

Theorem 5.10. *A nonidentity rotation fixes exactly one point, its center. A rotation with center C fixes every circle with center C. If C is a point and Θ and Φ are real numbers, then $\rho_{C,\Phi}\rho_{C,\Theta} = \rho_{C,\Theta+\Phi}$ and $\rho_{C,\Theta}^{-1} = \rho_{C,-\Theta}$. The rotations with center C form an abelian group. The involutory rotations are the halfturns, and $\rho_{C,180} = \sigma_C$ for any point C.*

Note that $\rho_{C,30} = \rho_{C,390} = \rho_{C,-330}$ as, in general, for real numbers Θ and Φ we have $\Theta° = \Phi°$ iff $\Theta = \Phi + k(360)$ for some integer k. Your conjecture involving rotations should also have involved a directed angle from line l to line m. For distinct intersecting lines l and m, there are *two* directed angles from l to m. Clearly, as in Figure 5.6, these will have directed angle measures that differ by a multiple of 180. (Fortunately, your conjecture should have involved *twice* a directed angle.) If Θ and Ψ are the directed angle measures of the two directed angles from l to m, then $(2\Theta)° = (2\Psi)°$ since numbers Θ and Ψ differ by a multiple of 180. For example, again with reference to Figure 5.6, we have $135 \neq -45$, $135° \neq -45°$, $270 \neq -90$, but $270° = -90°$. So, if we are talking about the rotation through twice a directed angle from line l to line m, then it makes no difference which of the two directed angles we choose.

Can you define the following key words and symbols introduced in this chapter: *congruent, rotation, $\rho_{P,\Theta}$, C_P*?

§5.3 Exercises

5.1.   Given $\triangle ABC \cong \triangle DEF$ where $A = (0, 0)$, $B = (5, 0)$, $C = (0, 10)$, $D = (4, 2)$, $E = (1, -2)$, and $F = (12, -4)$, find equations of lines such that the product of reflections in these lines takes $\triangle ABC$ to $\triangle DEF$.

5.2.   Suppose lines l, m, n have, respectively, equations $X = 2$, $Y = 3$, and $Y = 5$. Find the equations for $\sigma_m\sigma_l$ and for $\sigma_n\sigma_m$.

5.3.   Prove or disprove: Every isometry is either a product of five reflections or a product of six reflections.

5.4. Prove or disprove: The images of a triangle under two distinct isometries cannot be identical.

5.5. True or False
   (a) $(\sigma_z\sigma_y\sigma_x \cdots \sigma_c\sigma_b\sigma_a)^{-1} = \sigma_a\sigma_b\sigma_c \cdots \sigma_x\sigma_y\sigma_z$ for all lines $a, b, c, \ldots, x, y, z$.
   (b) If $A_B = C_D$, then $AD = BC$.
   (c) A product of four reflections is an isometry.
   (d) The set of all rotations generates an abelian group.
   (e) The set of all reflections generates $\mathscr{I}$.
   (f) If A and B are two distinct points, $PA = PB$, and $QA = QB$, then $P = Q$.
   (g) An isometry that fixes a point is an involution.
   (h) If isometry α fixes points A, B, and C, then $\alpha = \iota$.
   (i) If α and β are isometries and $\alpha^2 = \beta^2$, then $\alpha = \beta$ or $\alpha = \beta^{-1}$.
   (j) $\rho_{C,\theta}^{-1} = \rho_{C,-\theta} = \sigma_C$ for any point C.

5.6. Carry out the paper folding experiments described in the text. (If waxed paper is not available, try tracing paper.)

5.7. Give a reasonable definition for: $\square ABCD \cong \square SPQR$.

5.8. If $\square ABCD$ and $\square EFGH$ are congruent rectangles and $AB \neq BC$, then how many isometries are there that take one rectangle to the other?

5.9. Prove: If $\sigma_n\sigma_m$ fixes point P and $m \neq n$, then P is on both m and n.

5.10. Prove or disprove: If ρ is a rotation, then $\langle \rho \rangle$ is finite.

5.11. Prove or disprove: If T_1, T_2, T_3, T_4 are sets of points such that T_1 is congruent to T_2 while T_3 is congruent to T_4, then $T_1 \cap T_3$ is congruent to $T_2 \cap T_4$.

5.12. Prove or disprove: If α is an involution, then $\beta\alpha\beta^{-1}$ is an involution for any transformation β.

5.13. If the heads-up coin is rolled around the tails-up coin in Figure 5.7 until the heads-up coin is directly under the other, will the head then be upside down?

5.14. Given point P off line a, construct $\rho_{P,60}(a)$.

Figure 5.7

5.15. Given point A off two lines b and d, construct the squares $\square ABCD$ with B on b and D on d.

5.16. Given point P and two lines q and r, construct the equilateral triangles $\triangle PQR$ with Q on q and R on r.

5.17. Arrange the capital letters written in most symmetric form into equivalence classes where two letters are in the same class iff the two letters have the same symmetries when superimposed in standard orientation.

Chapter 6

The Product of Two Reflections

§6.1 Translations and Rotations

Every isometry is a product of at most three reflections (Theorem 5.6). So each isometry is of the form σ_l, $\sigma_m \sigma_l$, or $\sigma_r \sigma_m \sigma_l$. In this section the case $\sigma_m \sigma_l$ is examined. Since a reflection is an involution, we know $\sigma_l \sigma_l = \iota$ for any line l. Thus we are concerned with the product of two reflections in distinct lines l and m. There are two cases: either l and m are parallel or else l and m intersect at a unique point. We shall show first that if l and m are parallel lines then the product $\sigma_m \sigma_l$ is the translation through *twice* the directed distance from l to m.

Let l and m be distinct parallel lines. Suppose $\overleftrightarrow{LM}$ is a common perpendicular to l and m with L on l and M on m. The directed distance from l to m is the directed distance from L to M. We are going to use Theorem 5.2; look back at that theorem now. See Figure 6.1 below. With K a point on l distinct from L, let $L' = \sigma_m(L)$ and $K' = \tau_{L,L'}(K)$. Then (Theorems 3.1 and 3.2), we have $\tau_{K,K'} = \tau_{L,L'}$ and $\square LKK'L'$ is a rectangle with m the common perpendicular bisector of $\overline{LL'}$ and of $\overline{KK'}$. So $\sigma_m(K) = K'$. Now, let $J = \sigma_l(M)$. Then, since L is the midpoint of $\overline{JM}$ and M is the midpoint of $\overline{LL'}$, we have $\tau_{J,M} = \tau_{L,L'}$ where $\tau_{L,L'}$ is the translation through *twice* the directed distance from l to m. Hence,

$$\sigma_m \sigma_l(J) = \sigma_m(M) = M = \tau_{L,L'}(J),$$

$$\sigma_m \sigma_l(K) = \sigma_m(K) = K' = \tau_{L,L'}(K),$$

$$\sigma_m \sigma_l(L) = \sigma_m(L) = L' = \tau_{L,L'}(L).$$

Since an isometry is determined by any three noncollinear points (Theorem 5.2), the equations above give the desired result $\sigma_m \sigma_l = \tau_{L,L'} = \tau_{L,M}^2$.

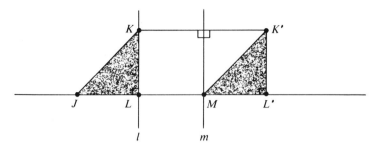

Figure 6.1

Theorem 6.1. *If lines l and m are parallel, then $\sigma_m \sigma_l$ is the translation through twice the directed distance from l to m.*

In the proof above we have $\tau_{L,L'} = \sigma_M \sigma_L$ (Theorem 3.6). Thus we have already proved the next theorem.

Theorem 6.2. *If line b is perpendicular to line l at L and to line m at M, then*

$$\sigma_m \sigma_l = \tau_{L,M}^2 = \sigma_M \sigma_L.$$

Is every translation a product of two reflections? Given nonidentity translation $\tau_{L,N}$, then $\tau_{L,N} = \sigma_M \sigma_L$ where M is the midpoint of $\overline{LN}$. With l the perpendicular to $\overleftrightarrow{LM}$ at L and m the perpendicular to $\overleftrightarrow{LM}$ at M, we have $\sigma_M \sigma_L = \sigma_m \sigma_l$ by the previous theorem. So $\tau_{L,N} = \sigma_m \sigma_l$ with $l \parallel m$. Thus we have proved a converse of our first theorem.

Theorem 6.3. *Every translation is a product of two reflections in parallel lines, and, conversely, a product of two reflections in parallel lines is a translation.*

The equations $\sigma_m \sigma_l = \sigma_n \sigma_p = \sigma_q \sigma_n$ have unique solutions for lines p and q when given lines l, m, n are parallel. To show this, let line b be perpendicular to parallel lines l, m, n at points L, M, N respectively. See Figure 6.2. Let P and Q be the unique points on b such that $\sigma_M \sigma_L = \sigma_N \sigma_P = \sigma_Q \sigma_N$. Let line p be perpendicular to b at P, and let line q be perpendicular to b at Q. Then

$$\sigma_m \sigma_l = \sigma_M \sigma_L = \sigma_N \sigma_P = \sigma_n \sigma_p \quad \text{and} \quad \sigma_m \sigma_l = \sigma_M \sigma_L = \sigma_Q \sigma_N = \sigma_q \sigma_n.$$

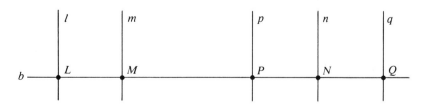

Figure 6.2

The uniqueness of these lines p and q that satisfy the equations follows from the cancellation laws; for example, $\sigma_n \sigma_p = \sigma_n \sigma_t$ implies $\sigma_p = \sigma_t$, which implies $p = t$. We have proved the following useful theorem.

Theorem 6.4. *If lines l, m, n are perpendicular to line b, then there are unique lines p and q such that*

$$\sigma_m \sigma_l = \sigma_n \sigma_p = \sigma_q \sigma_n.$$

Further, the lines p and q are perpendicular to b.

In the theorem above, note that p is just the unique line such that directed distance from p to n equals the directed distance from l to m and that q is just the unique line such that the directed distance from n to q equals the directed distance from l to m. Because of Theorem 6.3, we can restate the content of the important Theorem 6.4 as follows.

Theorem 6.5. *If $P \neq Q$, then $\tau_{P,Q}$ may be expressed as $\sigma_b \sigma_a$ where either one of a or b is an arbitrarily chosen line perpendicular to $\overleftrightarrow{PQ}$ and the other is then a uniquely determined line perpendicular to $\overleftrightarrow{PQ}$.*

Since $\sigma_m \sigma_l = \sigma_n \sigma_p$ and $\sigma_n \sigma_m \sigma_l = \sigma_p$ are seen to be equivalent equations by multiplying each by σ_n on the left, we can restate the content of Theorem 6.4 in still another form. The reflection in the statement of the next theorem is just σ_p of Theorem 6.4.

Theorem 6.6. *If lines l, m, n are perpendicular to line b, then $\sigma_n \sigma_m \sigma_l$ is a reflection in a line perpendicular to b.*

We now turn to the case where l and m are distinct lines intersecting at a point C. We shall follow much the same path as we did for parallel lines. We first show $\sigma_m \sigma_l$ is a rotation about C by using the theorem that three non-collinear points determine an isometry. Suppose $\Theta/2$ is the directed angle measure of one of the two directed angles from l to m. We may as well suppose $-90 < \Theta/2 \leq 90$. Note that the notation suggests correctly that we are going to encounter *twice* the directed angle from l to m in our conclusion. Let L be a point on l different from C. Let point M be the intersection of line m and circle C_L such that the directed angle measure from $\overrightarrow{CL}$ to $\overrightarrow{CM}$ is $\Theta/2$. See Figure 6.3. We have $l = \overleftrightarrow{CL}$ and $m = \overleftrightarrow{CM}$. Let $L' = \rho_{C,\Theta}(L)$. Then L' is on circle C_L, and m is the perpendicular bisector of $\overline{LL'}$. So $L' = \sigma_m(L)$. Let $J = \sigma_l(M)$. Then l is the perpendicular bisector of $\overline{JM}$. So J is on circle C_L, and the directed angle measure from $\overrightarrow{CJ}$ to $\overrightarrow{CM}$ is Θ. Hence $M = \rho_{C,\Theta}(J)$. Therefore,

$$\sigma_m \sigma_l(C) = \sigma_m(C) = C = \rho_{C,\Theta}(C),$$

$$\sigma_m \sigma_l(J) = \sigma_m(M) = M = \rho_{C,\Theta}(J),$$

$$\sigma_m \sigma_l(L) = \sigma_m(L) = L' = \rho_{C,\Theta}(L).$$

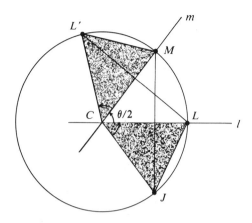

Figure 6.3

Since points C, J, L are not collinear, we conclude $\sigma_m\sigma_l = \rho_{C,\Theta}$. So $\sigma_m\sigma_l$ is the rotation about C through *twice* a directed angle from l to m. The discussion at the end of Chapter 5 explains why either directed angle from l to m may be used in the statement of our theorem.

Theorem 6.7. *If lines l and m intersect at point C and the directed angle measure of a directed angle from l to m is $\Theta/2$, then $\sigma_m\sigma_l = \rho_{C,\Theta}$.*

Conversely, suppose $\rho_{C,\Theta}$ is given. Let l be any line through C, and let m be the line through C such that a directed angle from l to m has directed angle measure $\Theta/2$. Then $\rho_{C,\Theta} = \sigma_m\sigma_l$, and we have the following analogue to Theorem 6.3.

Theorem 6.8. *Every rotation is a product of two reflections in intersecting lines, and, conversely, a product of two reflections in intersecting lines is a rotation.*

Given rays $\overrightarrow{CL}$, $\overrightarrow{CM}$, and $\overrightarrow{CN}$, there are unique rays $\overrightarrow{CP}$ and $\overrightarrow{CQ}$ such that the directed angle from $\overrightarrow{CL}$ to $\overrightarrow{CM}$, the directed angle from $\overrightarrow{CP}$ to $\overrightarrow{CN}$, and the directed angle from $\overrightarrow{CN}$ to $\overrightarrow{CQ}$ all have the same directed angle measure. See Figure 6.4. With $n = \overleftrightarrow{CN}$, $p = \overleftrightarrow{CP}$, and $q = \overleftrightarrow{CQ}$, we have solutions p and q to the equations $\sigma_m\sigma_l = \sigma_n\sigma_p = \sigma_q\sigma_n$ when l, m, n are given lines concurrent at C. The uniqueness of such lines p and q follows from the cancellation laws. Because of its importance, our result is stated in three different ways, analogous to Theorems 6.4, 6.5, and 6.6.

Theorem 6.9. *If lines l, m, n are concurrent at point C, then there are unique lines p and q such that*

$$\sigma_m\sigma_l = \sigma_n\sigma_p = \sigma_q\sigma_n.$$

Further, the lines p and q are concurrent at C.

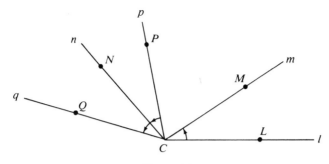

Figure 6.4

Theorem 6.10. *Rotation $\rho_{C,\Theta}$ may be expressed as $\sigma_b \sigma_a$ where either one of a or b is an arbitrarily chosen line through C and the other is then a uniquely determined line through C.*

Theorem 6.11. *If lines l, m, n are concurrent at point C, then $\sigma_n \sigma_m \sigma_l$ is a reflection in a line through C.*

Since $\rho_{P,180} = \sigma_P$ for any point P, a special case of Theorem 6.10 has the following form.

Theorem 6.12. *Halfturn σ_p is the product (in either order) of the two reflections in any two lines perpendicular at P.*

Of course, $\sigma_l \sigma_l = \tau_{P,P} = \rho_{P,0} = \iota$ for any line l and any point P. Also, a rotation has a fixed point while a nonidentity translation does not. From these observations and the fact that lines l and m must be parallel or intersect, we have the last theorem of this section.

Theorem 6.13. *A product of two reflections is a translation or a rotation; only the identity is both a translation and a rotation.*

§6.2 Fixed Points and Involutions

We have not considered products of three reflections, except in the very special cases where the reflections are in lines that are parallel or in lines that are concurrent. Therefore, it would be fairly surprising if we could at this stage classify all the isometries that have fixed points and classify all the isometries that are involutions. Such is the case, however.

An isometry with a fixed point is a product of at most two reflections (Theorem 5.5). Of course, the identity and a reflection have fixed points. Otherwise, an isometry with fixed points must be a translation or a rotation

(Theorem 6.13). Since a nonidentity translation has no fixed points and a nonidentity rotation has exactly one fixed point, we have the following classification of the isometries with fixed points.

Theorem 6.14. *An isometry that fixes exactly one point is a rotation. An isometry that fixes a point is a rotation or a reflection.*

The involutions come next. Suppose α is an involutory isometry. Since α is not the identity, there are points P and Q such that $\alpha(P) = Q \neq P$. Since $P = \alpha^2(P) = \alpha(Q)$, then α interchanges distinct points P and Q. Hence (Theorem 4.4), α must fix the midpoint of $\overline{PQ}$. Therefore α must be a rotation or a reflection by the previous theorem. Since the involutory rotations are the halfturns (Theorem 5.10), we have the following classification of the isometries that are involutions.

Theorem 6.15. *The involutory isometries are the reflections and the halfturns.*

Although we know (Theorem 3.5) that halfturn σ_P fixes line l iff point P is on line l, we have not considered the fixed lines of an arbitrary rotation. We do so now, and suppose nonidentity rotation $\rho_{C,\Theta}$ fixes line l. Let m be the line through C that is perpendicular to l. Then (Theorem 6.10), there is a line n through C and different from m such that $\rho_{C,\Theta} = \sigma_n\sigma_m$. Since l and m are perpendicular, then (Theorem 4.1) we have $l = \rho_{C,\Theta}(l) = \sigma_n\sigma_m(l) = \sigma_n(l)$. So σ_n fixes line l. Then, $n = l$ or $n \perp l$. Lines m and n cannot be two intersecting lines and both perpendicular to l. Hence, $n = l$. So m and n are perpendicular at C and $\rho_{C,\Theta}$ is the halfturn σ_C.

Theorem 6.16. *A nonidentity rotation that fixes a line is a halfturn.*

The useful trick of interchanging "$\sigma_n\sigma_l\sigma_l\sigma_m$" and "$\sigma_n\sigma_m$" is akin in arithmetic to multiply or dividing by 1 in some clever form, say to reduce a fraction or to rationalize the denominator of the fraction. The advantage of factoring out the identity $\sigma_l\sigma_l$ is to simplify. However, sometimes it is convenient to insert the identity as a factor. For example, if $m \parallel n$, the quickest way to find points M and N such that $\sigma_n\sigma_m = \sigma_N\sigma_M$ is to insert the identity in the form "$\sigma_l\sigma_l$" where l is conveniently chosen to be a common perpendicular to m and n, say at points M and N, respectively. Then

$$\sigma_n\sigma_m = \sigma_n l \sigma_m = \sigma_n(\sigma_l\sigma_l)\sigma_m = (\sigma_n\sigma_l)(\sigma_l\sigma_m) = \sigma_N\sigma_M.$$

Reading the line above backwards we see how to solve the converse problem of getting from $\sigma_N\sigma_M$ to $\sigma_n\sigma_m$. Similar to that problem is the problem of finding the fixed point of $\rho_{B,90}\rho_{A,60}$ in Figure 6.5. The line l that relates these rotations is the line through the centers A and B. There is a line m' through A such that $\rho_{A,60} = \sigma_{m'}\sigma_l$. True, but we are not interested. We want the common "σ_l" to be in the middle of our product. There is a line m through A

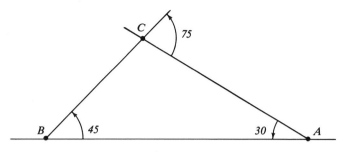

Figure 6.5

such that $\rho_{A,\,60} = \sigma_l\sigma_m$ and there is a line n through B such that $\rho_{B,\,90} = \sigma_n\sigma_l$. Then it is no surprise that

$$\rho_{B,\,90}\,\rho_{A,\,60} = \sigma_n\sigma_l\sigma_l\sigma_m = \sigma_n\sigma_m = \rho_{C,\,150}.$$

We finish this section with a study of the symmetry of the alphabet. The problem (Exercise 5.17) is to arrange the capital letters *written in most symmetric form* into equivalence classes where two letters are in the same class iff the letters have the same symmetries when superimposed in standard orientation. Do you agree that there are the ten classes given by the columns of Table 6.1? An interesting word game is to try to make up a ten-letter

Table 6.1

A	B	F	H	L	N	O	Q	X	Y
M	C	G	I		S				
T	D	J			Z				
U	E	R							
V	K	P							
W									

word or phrase that uses exactly one letter from each of the ten classes. For example, see Figure 6.6.

Figure 6.6

§6.3 Exercises

6.1. Prove: If P is a point and l and m are lines, then there are lines p and q such that P is on p and $\sigma_m \sigma_l = \sigma_q \sigma_p$.

6.2. In Figure 6.7, sketch the fixed point of $\sigma_t \sigma_s$ where $\sigma_s = \sigma_n \sigma_m \sigma_l$ and $\sigma_t = \sigma_c \sigma_b \sigma_a$.

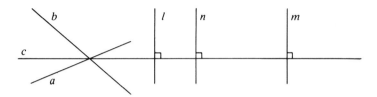

Figure 6.7

6.3. Given a figure consisting of two points P and Q, sketch a construction of the fixed point of $\tau_{P,Q} \rho_{P,45}$.

6.4. Given a figure consisting of three points A, B, C, sketch a construction of the fixed point of $\tau_{B,C} \rho_{A,120}$.

6.5. What are the equations for $\sigma_n \sigma_m$ if line m has equation $Y = -2X + 3$ and line n has equation $Y = -2X + 8$?

6.6. Show that $\sigma_l \rho_{C,\Theta} \sigma_l = \rho_{C,-\Theta}$ if point C is on line l.

6.7. True or False
   (a) If a directed angle from line l to line m is $240°$, then $\sigma_m \sigma_l$ is a rotation of $120°$.
   (b) $\sigma_m \sigma_l = \tau_{L,M}^2 = \sigma_M \sigma_L$ if point L is on line l and point M is on line m.
   (c) An isometry has a unique fixed point iff the isometry is a nonidentity rotation.
   (d) An isometry that is its own inverse must be a halfturn, a reflection, or the identity.
   (e) If $L' = \sigma_m(L)$ and $K' = \tau_{L,L'}(K)$, then m is the perpendicular bisector of $\overline{KK'}$.
   (f) Given points L, M, N, there is a point P such that $\sigma_M \sigma_L = \sigma_N \sigma_P$.
   (g) Given lines l, m, n, there is a line p such that $\sigma_m \sigma_l = \sigma_n \sigma_p$.
   (h) If lines l and m intersect at point C and a directed angle from l to m is $\Phi°$, then $\sigma_m \sigma_l = \rho_{C,2\Phi}$.
   (i) An isometry that fixes a point must be a rotation, a reflection, or the identity.
   (j) Isometry $\alpha \beta \alpha^{-1}$ is an involution for any isometry α iff isometry β is an involution.

6.8. Given nonparallel lines $\overleftrightarrow{AB}$ and $\overleftrightarrow{CD}$, show there is a rotation ρ such that $\rho(\overrightarrow{AB}) = \overrightarrow{CD}$.

6.9. Prove or disprove: Every translation is a product of two noninvolutory rotations.

6.10. Prove or disprove: If $P \neq Q$, then there is a unique translation taking point P to point Q but there are an infinite number of rotations that take P to Q.

6.11. If l, m, n are the perpendicular bisectors of sides $\overline{AB}$, $\overline{BC}$, $\overline{CA}$, respectively, of $\triangle ABC$, then $\sigma_n \sigma_m \sigma_l$ is a reflection in which line?

6.12. What lines are fixed by rotation $\rho_{C,\Theta}$?

6.13. Prove Theorem 6.4 follows from Theorem 6.5.

6.14. If $\sigma_c \sigma_b \sigma_a$ is a reflection, show that lines a, b, c are either concurrent or parallel to each other.

6.15. Show that $\sigma_n \sigma_m \sigma_l = \sigma_l \sigma_m \sigma_n$ whenever lines l, m, n are concurrent or have a common perpendicular.

6.16. Show that the product of the reflections in the three angle bisectors of a triangle is a reflection in a line perpendicular to a side of the triangle.

6.17. If $\sigma_n \sigma_m((x, y)) = (x + 6, y - 3)$, find equations for lines m and n.

6.18. If l and m are distinct intersecting lines, find the locus of all points P such that $\rho_{P,\Theta}(l) = m$ for some Θ.

Chapter 7

Even Isometries

§7.1 Parity

A product of two reflections is a translation or a rotation. By considering the fixed points of each, we see that neither a translation nor a rotation can be equal to a reflection. Thus, $\sigma_n \sigma_m \neq \sigma_l$ for lines l, m, n. When a given isometry is expressed as a product of reflections, the number of reflections is not invariant; indeed, we can always add 2 to the number of reflections in a given product by inserting "$\sigma_l \sigma_l$" into the product. Although a product of two reflections cannot be a reflection, we know that in some cases a product of three reflections is a reflection. We shall see this is possible only because both 3 and 1 are odd integers. In mathematics, *parity* refers to the property of an integer being either even or odd. An isometry that is a product of an even number of reflections is said to be *even*; an isometry that is a product of an odd number of reflections is said to be *odd*. Since an isometry is a product of reflections, then an isometry is even or odd. However, the definition will be useful only if we show that no isometry is both even and odd. Of course no integer can be both even and odd, but is it not conceivable some product of ten reflections could be equal to some product of seven reflections? To show this is impossible, we first show that a product of four reflections is always equal to a product of two reflections.

Our argument that a product of four reflections is a product of two reflections depends on two applications of the lemma: If P is a point and a and b are lines, then there are lines c and d with c passing through P such that $\sigma_b \sigma_a = \sigma_d \sigma_c$. Suppose a, b, P are given. If $a \parallel b$, let c be the line through P that is parallel to a; if a and b intersect at C, let c be a line through P and C. So a, b, c are either parallel to each other or else concurrent. In either case, there is a line d such that $\sigma_b \sigma_a \sigma_c = \sigma_d$. Then $\sigma_b \sigma_a = \sigma_d \sigma_c$ and we have the

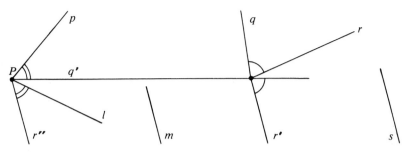

Figure 7.1

desired lemma. Now, suppose product $\sigma_s \sigma_r \sigma_q \sigma_p$ is given. We want to show this product is equal to a product of two reflections. Let P be a point on line p. See Figure 7.1 for an example. By the lemma, there are lines q' and r' such that $\sigma_r \sigma_q = \sigma_{r'} \sigma_{q'}$ with P on q'. Again by the lemma, there are lines r'' and m such that $\sigma_s \sigma_{r'} = \sigma_m \sigma_{r''}$ with P on r''. Since p, q', r'' are concurrent at P, then there is a line l such that $\sigma_{r''} \sigma_{q'} \sigma_p = \sigma_l$. Therefore,

$$\sigma_s \sigma_r \sigma_q \sigma_p = \sigma_s \sigma_{r'} \sigma_{q'} \sigma_p = \sigma_m \sigma_{r''} \sigma_{q'} \sigma_p = \sigma_m \sigma_l.$$

Not only are there lines such that the given product of four reflections is equal to $\sigma_m \sigma_l$, but our proof even tells how to find such lines.

Theorem 7.1. *A product of four reflections is a product of two reflections.*

Given a long product of reflections, we can use this theorem repeatedly to replace the first four reflections by two reflections until we have obtained a product with less than four reflections. By repeated application of the theorem to an even isometry, we can reduce the even isometry to a product of two reflections. Also, by repeated application of the theorem to an odd isometry, we can reduce the odd isometry to a product of three reflections or to a reflection. Therefore, to show an isometry cannot be both even and odd, we need to show only that a product of two reflections cannot equal a reflection or a product of three reflections. Assume there are lines p, q, r, s, t such that $\sigma_r \sigma_q \sigma_p = \sigma_s \sigma_t$. Then, we have shown above that there are lines l and m such that $\sigma_m \sigma_l = \sigma_s \sigma_r \sigma_q \sigma_p = \sigma_s \sigma_s \sigma_t = \sigma_t$. We have a contradiction since $\sigma_m \sigma_l$ is a translation or a rotation and cannot be equal to reflection σ_t. A product of two reflections is never equal to a reflection or a product of three reflections. We have our desired theorem concerning parity.

Theorem 7.2. *An even isometry is a product of two reflections. An odd isometry is a reflection or a product of three reflections. No isometry is both even and odd.*

The even isometries are the translations and the rotations. Since the involutory isometries are the halfturns and the reflections, the theorem above gives the following partition of these involutions.

Theorem 7.3. *An even involutory isometry is a halfturn; an odd involutory isometry is a reflection.*

An isometry and its inverse have the same parity, since the inverse of a product of reflections is the product of the reflections in reverse order. So the set $\mathscr{E}$ of all even isometries has the inverse property. Further, the set $\mathscr{E}$ has the closure property since the sum of two even integers is even. That's all we need to show that $\mathscr{E}$ forms a group.

Theorem 7.4. *The even isometries form a group $\mathscr{E}$.*

$\mathscr{E}$ will always denote the group of even isometries. So $\mathscr{E}$ consists of the translations and the rotations. Since some persons call only these isometries in $\mathscr{E}$ *motions* while others call any isometry in $\mathscr{I}$ a *motion*, we shall get along without using the term at all.

$\alpha\sigma_P\alpha^{-1} = \sigma_{\alpha(P)}$ for any isometry α since $\alpha\sigma_P\alpha^{-1}$ is an even involutory isometry that fixes $\alpha(P)$. That's true but a little fast. Suppose α and β are isometries. Then, $\alpha\beta\alpha^{-1}$ is an involution iff β is an involution since $(\alpha\beta\alpha^{-1})^2 = \alpha\beta^2\alpha^{-1} = \iota$ iff $\beta^2 = \iota$ and since $\alpha\beta\alpha^{-1} = \iota$ iff $\beta = \iota$. For example, since σ_P and σ_m are involutions, then $\alpha\sigma_P\alpha^{-1}$ and $\alpha\sigma_m\alpha^{-1}$ are involutions. Further, $\alpha\beta\alpha^{-1}$ and β must have the same parity since α and its inverse α^{-1} have the same parity. For example, $\alpha\sigma_P\alpha^{-1}$ is even because σ_P is even, and $\alpha\sigma_m\alpha^{-1}$ is odd because σ_m is odd. Since $\alpha\sigma_P\alpha^{-1}$ is an even involution, then $\alpha\sigma_P\alpha^{-1}$ must be a halfturn (Theorem 7.3). Hence, since halfturn $\alpha\sigma_P\alpha^{-1}$ clearly fixes point $\alpha(P)$, then $\alpha\sigma_P\alpha^{-1}$ must be the halfturn about $\alpha(P)$. That is, $\alpha\sigma_P\alpha^{-1} = \sigma_{\alpha(P)}$. In similar fashion, since $\alpha\sigma_m\alpha^{-1}$ is an odd involutory isometry, then $\alpha\sigma_m\alpha^{-1}$ is a reflection. Hence, since $\alpha\sigma_m\alpha^{-1}$ clearly fixes every point $\alpha(P)$ on line $\alpha(m)$, then $\alpha\sigma_m\alpha^{-1}$ must be the reflection in the line $\alpha(m)$. That is, $\alpha\sigma_m\alpha^{-1} = \sigma_{\alpha(m)}$.

Theorem 7.5. *If P is a point, m is a line, and α is an isometry, then*

$$\alpha\sigma_m\alpha^{-1} = \sigma_{\alpha(m)} \quad and \quad \alpha\sigma_P\alpha^{-1} = \sigma_{\alpha(P)}.$$

Figure 7.2 illustrates an example of the first equation in the theorem with α taken to be $\rho_{P, 60}$. The theorem says that if we rotate the points of the plane about P through $-60°$, then reflect in line m, and then rotate about P through $+60°$, the effect of all this is just the same as only reflecting in line n where $n = \rho_{P, 60}(m)$. That seems truly amazing.

In general, $\alpha\beta\alpha^{-1}$ is called the **conjugate** of β by α. We next look at the conjugate $\alpha\tau_{A, B}\alpha^{-1}$ of translation $\tau_{A, B}$ by isometry α. If M is the midpoint of A and B, then point $\alpha(M)$ is the midpoint of points $\alpha(A)$ and $\alpha(B)$. Also, $\tau_{A, B} = \sigma_M\sigma_A$ and $\tau_{\alpha(A), \alpha(B)} = \sigma_{\alpha(M)}\sigma_{\alpha(A)}$. Using our method of inserting the identity in a useful form, we then have

$$\alpha\tau_{A, B}\alpha^{-1} = \alpha\sigma_M\sigma_A\alpha^{-1} = \alpha\sigma_M\alpha^{-1}\alpha\sigma_A\alpha^{-1} = \sigma_{\alpha(M)}\sigma_{\alpha(A)} = \tau_{\alpha(A), \alpha(B)}.$$

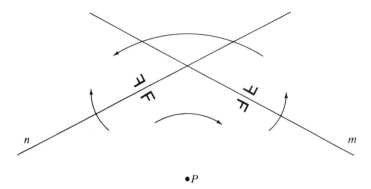

Figure 7.2

What could be nicer than the result $\alpha\tau_{A,B}\alpha^{-1} = \tau_{\alpha(A),\alpha(B)}$? Finding the conjugate of a rotation is slightly more complicated. We first examine the conjugate of $\rho_{C,\Theta}$ by σ_l. Let m be the line through C that is perpendicular to l. Then there is a line n through C such that $\rho_{C,\Theta} = \sigma_n\sigma_m$. See Figure 7.3. Now $\sigma_l(m)$ and $\sigma_l(n)$ intersect at $\sigma_l(C)$, and a directed angle from $\sigma_l(m)$ to $\sigma_l(n)$ is the negative of a directed angle from m to n. This explains the negative sign on the far right in the following calculation:

$$\sigma_l\rho_{C,\Theta}\sigma_l^{-1} = \sigma_l\sigma_n\sigma_m\sigma_l^{-1} = \sigma_l\sigma_n\sigma_l^{-1}\sigma_l\sigma_m\sigma_l^{-1} = \sigma_{\sigma_l(n)}\sigma_{\sigma_l(m)} = \rho_{\sigma_l(C),-\Theta}.$$

If $\alpha = \sigma_t\sigma_s$, then $\alpha\rho_{C,\Theta}\alpha^{-1} = \sigma_t(\sigma_s\rho_{C,\Theta}\sigma_s^{-1})\sigma_t^{-1} = \rho_{\alpha(C),+\Theta}$ with a positive sign replacing two negative signs in front of Θ. If $\alpha = \sigma_t\sigma_s\sigma_r$, then the sign in front of Θ is back to a negative sign again. We summarize our results as follows.

Theorem 7.6. *If α is an isometry, then*

$$\alpha\tau_{A,B}\alpha^{-1} = \tau_{\alpha(A),\alpha(B)} \quad and \quad \alpha\rho_{C,\Theta}\alpha^{-1} = \rho_{\alpha(C),\pm\Theta},$$

where the positive sign applies when α is even and the negative sign applies when α is odd.

By taking $\alpha = \rho_{D,\Phi}$ in the theorem, we can show nonidentity rotation $\rho_{D,\Phi}$ does not commute with nonidentity rotation $\rho_{C,\Theta}$ unless $D = C$. We

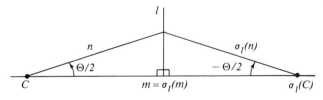

Figure 7.3

state this as a theorem here but leave the details of the argument as Exercise 7.1.

Theorem 7.7. *Nonidentity rotations with different centers do not commute.*

We are also in a position to answer the question, "When do reflections commute?" For any lines m and n, the following five statements are seen to be equivalent. (1) $\sigma_m \sigma_n = \sigma_n \sigma_m$. (2) $\sigma_n \sigma_m \sigma_n = \sigma_m$. (3) $\sigma_{\sigma_n(m)} = \sigma_m$. (4) $\sigma_n(m) = m$. (5) $m = n$ or $m \perp n$. Comparing (1) and (5), we have the answer to our question.

Theorem 7.8. $\sigma_m \sigma_n = \sigma_n \sigma_m$ *iff* $m = n$ *or* $m \perp n$.

Several equivalences between algebraic equations and geometric relations similar to that in Theorem 7.8 are indicated in Exercise 7.3. These are verified in the same manner.

We now consider products of even isometries. We already know (Theorem 3.4) that the product of two translations is a translation. We also know (Theorem 3.6) that the product of two rotations can be a translation in some cases. For example, $\sigma_B \sigma_A = \tau_{A,B}^2$; don't forget that a halfturn is a rotation of 180°. We know $\rho_{C,\Phi} \rho_{C,\Theta} = \rho_{C,\Theta+\Phi}$ by Theorem 5.10. Let's consider the product $\rho_{B,\Phi} \rho_{A,\Theta}$ of two nonidentity rotations with different centers. With $c = \overleftrightarrow{AB}$, there is a line a through A and a line b through B such that $\rho_{A,\Theta} = \sigma_c \sigma_a$ and $\rho_{B,\Phi} = \sigma_b \sigma_c$. So $\rho_{B,\Phi} \rho_{A,\Theta} = \sigma_b \sigma_c \sigma_c \sigma_a = \sigma_b \sigma_a$. When $(\Theta + \Phi)° = 0°$, then the lines a and b are parallel and our product is a translation. This is easier to see when the directed angles are chosen as interior angles on the same side of c; for example, in Figure 7.4 we have $\Theta/2 + \Phi/2 = -180$. On the other hand, when $(\Theta + \Phi)° \neq 0°$, then the lines a and b intersect at some point C and our product is a rotation.

More than that, with the directed angles chosen as interior angles on the same side of c as C, we can see by the Exterior Angle Theorem from elementary geometry that one directed angle from a to b is $(\Theta/2 + \Phi/2)°$. See Figure 7.5. Hence our product $\sigma_b \sigma_a$ is a rotation about C through an angle of $(\Theta + \Phi)°$. That is, $\rho_{B,\Phi} \rho_{A,\Theta} = \rho_{C,\Theta+\Phi}$. Now, what is the product of a translation

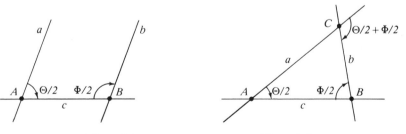

Figure 7.4   Figure 7.5

and a nonidentity rotation? In such a product we replace the translation by a product of two rotations of 180° each and obtain as the product of the three rotations a rotation of $(\Theta + 180 + 180)°$, which is just a rotation of $\Theta°$. We have proved *The Angle-addition Theorem.*

Theorem 7.9. *A rotation of $\Theta°$ followed by a rotation of $\Phi°$ is a rotation of $(\Theta + \Phi)°$ unless $(\Theta + \Phi)° = 0°$, in which case the product is a translation. A translation followed by a nonidentity rotation of $\Theta°$ is a rotation of $\Theta°$. A nonidentity rotation of $\Theta°$ followed by a translation is a rotation of $\Theta°$. A translation followed by a translation is a translation.*

The Angle-addition Theorem can also be proved by using the equations for the even isometries that will be developed in Section 9.1.

§7.2 The Dihedral Groups

We are going to compute the symmetry group of a regular polygon. As a specific example of what we are going to do, we first consider the square. We suppose the square is centered at the origin in the Cartesian plane and that one vertex lies on the positive X-axis. With the notation as in Figure 7.6, we see that the square is fixed by ρ and by σ where $\rho = \rho_{O, 90}$ and $\sigma = \sigma_h$. Note that $\rho^4 = \sigma^2 = \iota$. Since the symmetries of the square form a group, then the square must be fixed by the four distinct rotations ρ, ρ^2, ρ^3, ρ^4 and by the four distinct odd isometries $\rho\sigma$, $\rho^2\sigma$, $\rho^3\sigma$, $\rho^4\sigma$. Let V_1 and V_2 be adjacent vertices of the square. Under a symmetry, V_1 may go to any one of the four vertices, but then V_2 must go to one of the two vertices adjacent to that one and the images of all remaining vertices are then determined. So

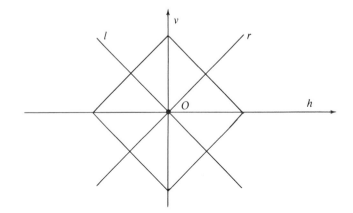

Figure 7.6

there are at most eight symmetries for the square. We have listed eight distinct symmetries above. Therefore, there are exactly eight symmetries and we have listed all of them. Isometries ρ and σ generate the entire group. The symmetry group of the square is $\langle \rho, \sigma \rangle$, a group of order 8 and denoted by D_4. Check each of the following calculations.

$$\rho = \sigma_r \sigma_h = \rho^{-3} \qquad \rho\sigma = \sigma_r \qquad \sigma\rho = \rho^3 \sigma$$

$$\rho^2 = \sigma_v \sigma_h = \rho^{-2} \qquad \rho^2\sigma = \sigma_v \qquad \sigma\rho^2 = \rho^2 \sigma$$

$$\rho^3 = \sigma_l \sigma_h = \rho^{-1} \qquad \rho^3\sigma = \sigma_l \qquad \sigma\rho^3 = \rho\sigma$$

The Cayley table for D_4 is given in Table 7.1. Let's see how to compute this table. The first row is trivial. The ρ row is easily computed by just multiplying each corresponding element in the ι row by ρ on the left. Likewise, the ρ^2 row is obtained by multiplying each corresponding element in the ρ row by ρ on the left, and the ρ^3 row is obtained by multiplying each corresponding element in the ρ^2 row by ρ on the left. Now the σ row is quite different. The equations in the third column above tell how a σ "hops over a power of ρ." Once we have computed the σ row, we return to computing each of the remaining rows by multiplying each corresponding element in the row above by ρ on the left. So the whole table is easy to compute once the σ row is known. However, since O is on the X-axis, then $\sigma\rho^k$ is a reflection in a line through O (Theorem 6.11), and hence an involution. So $\sigma\rho^k = (\sigma\rho^k)^{-1} = \rho^{-k}\sigma = \rho^{4-k}\sigma$. Therefore, to compute the entire Cayley table, all that is needed are the equations $\sigma\rho^k = \rho^{-k}\sigma$ and $\rho^4 = \sigma^2 = \iota$. Study this special case involving the square before going on to the general regular polygon.

Let n be a positive integer greater than 2. Suppose a regular n-gon is centered at the origin in the Cartesian plane and that one vertex lies on the positive X-axis. The n-gon is fixed by ρ and by σ where $\rho = \rho_{O,\,360/n}$ and σ is the reflection in the X-axis. Note that $\rho^n = \sigma^2 = \iota$. Since the symmetries of the n-gon form a group, then the n-gon must be fixed by the n distinct even rotations $\rho, \rho^2, \ldots, \rho^n$ and by the n distinct odd isometries $\rho\sigma, \rho^2\sigma, \ldots, \rho^n\sigma$.

Table 7.1

D_4	ι	ρ	ρ^2	ρ^3	σ	$\rho\sigma$	$\rho^2\sigma$	$\rho^3\sigma$
ι	ι	ρ	ρ^2	ρ^3	σ	$\rho\sigma$	$\rho^2\sigma$	$\rho^3\sigma$
ρ	ρ	ρ^2	ρ^3	ι	$\rho\sigma$	$\rho^2\sigma$	$\rho^3\sigma$	σ
ρ^2	ρ^2	ρ^3	ι	ρ	$\rho^2\sigma$	$\rho^3\sigma$	σ	$\rho\sigma$
ρ^3	ρ^3	ι	ρ	ρ^2	$\rho^3\sigma$	σ	$\rho\sigma$	$\rho^2\sigma$
σ	σ	$\rho^3\sigma$	$\rho^2\sigma$	$\rho\sigma$	ι	ρ^3	ρ^2	ρ
$\rho\sigma$	$\rho\sigma$	σ	$\rho^3\sigma$	$\rho^2\sigma$	ρ	ι	ρ^3	ρ^2
$\rho^2\sigma$	$\rho^2\sigma$	$\rho\sigma$	σ	$\rho^3\sigma$	ρ^2	ρ	ι	ρ^3
$\rho^3\sigma$	$\rho^3\sigma$	$\rho^2\sigma$	$\rho\sigma$	σ	ρ^3	ρ^2	ρ	ι

The symmetry group of the n-gon must have at least these $2n$ symmetries. Let V_1 and V_2 be adjacent vertices of the n-gon. Under a symmetry, V_1 may go to any of the n vertices, but then V_2 must go to one of the two vertices adjacent to that one and the images of all remaining vertices are then determined. So there are at most $2n$ symmetries for the n-gon. Therefore, there are exactly $2n$ symmetries of the n-gon and we have listed all of them. Isometries ρ and σ generate the entire group. The symmetry group of the n-gon is $\langle \rho, \sigma \rangle$, a group of order $2n$ denoted by D_n. These symmetry groups D_n are called ***dihedral groups***, as are groups D_1 and D_2 where $D_1 = \langle \sigma \rangle$ and $D_2 = \langle \rho, \sigma \rangle$ with ρ a rotation about O of $180°$. Since O is on the X-axis, then $\sigma \rho^k$ is a reflection in a line through O (Theorem 6.11) and hence is an involution for any integer k. Therefore, $\sigma \rho^k = (\sigma \rho^k)^{-1} = \rho^{-k}\sigma = \rho^{n-k}\sigma$. (We can also obtain this equation from Theorem 7.6 with $\alpha = \sigma$ and $\rho_{C,\Theta} = \rho^k$.) With the elements of D_n written in the form $\iota, \rho, \rho^2, \ldots, \rho^{n-1}, \sigma, \rho\sigma, \rho^2\sigma, \ldots, \rho^{n-1}\sigma$, we can easily write down the entire Cayley table for D_n just by using the equations $\sigma \rho^k = \rho^{-k}\sigma$ and $\rho^n = \sigma^2 = \iota$.

Groups D_1 and D_2 are, respectively, symmetry groups of an isosceles triangle that is not equilateral and of a rectangle that is not a square. (Note that D_2 is the familiar group V_4 from Section 2.2.) For any positive integer n, the subgroup of D_n containing all the even isometries in D_n is denoted by C_n. So C_n is the cyclic group of order n generated by ρ where $\rho = \rho_{O, 360/n}$. Group C_1 contains only the identity element and is the symmetry group of a scalene triangle. Since C_2 contains only the identity and a halfturn, then C_2 is the symmetry group of a parallelogram that is not a rhombus. For $n > 2$, group C_n is the symmetry group of a $2n$-gon akin to that in Figure 4.5 for $n = 3$; the trick is to take a properly chosen fourth of each side of a regular n-gon to be an alternating side of the $2n$-gon. The case for $n = 6$ and a ratchet polygon also having symmetry group C_6 are shown in Figure 7.7. We have defined the groups C_n and D_n and verified the following theorem.

Theorem 7.10. *For each positive integer n, there is a polygon having symmetry group D_n and a polygon having symmetry group C_n.*

Do you suppose that, conversely, every polygon has a symmetry group C_n or D_n for some n? This will be discussed in the next chapter.

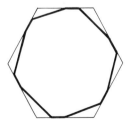

Figure 7.7

Can you define the following key words and symbols introduced in this chapter: *even, odd, &, C_n, D_n*?

§7.3 Exercises

7.1. Prove: Nonidentity rotations with different centers do not commute (Theorem 7.7).

7.2. In Figure 7.8, sketch lines l and m such that $\sigma_s \sigma_r \sigma_q \sigma_p = \sigma_m \sigma_l$.

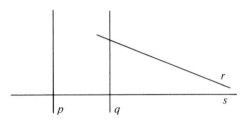

Figure 7.8

7.3. Express each of the following equations in a form that does not involve isometries:

$$\sigma_A \sigma_B = \sigma_B \sigma_A, \qquad \sigma_B \sigma_A = \sigma_C \sigma_B, \qquad \sigma_B \sigma_a = \sigma_b \sigma_B,$$

$$\sigma_b \sigma_A = \sigma_B \sigma_b, \qquad \sigma_b \sigma_A = \sigma_A \sigma_b, \qquad \sigma_b \sigma_a = \sigma_c \sigma_b.$$

7.4. Compute a Cayley table for D_3.

7.5. Sketch a polygon having symmetry group C_5.

7.6. Find all points that have the same image under each of two given rotations.

7.7. True or False
   (a) An even isometry that fixes two points is the identity.
   (b) The set of rotations generates $\mathscr{E}$.
   (c) An odd isometry is a product of three reflections.
   (d) An even isometry is a product of four reflections.
   (e) If $\rho_{\alpha(C),\Theta} = \rho_{C,\Theta}$ for isometry α, then α fixes C.
   (f) For any isometry α, any points A, B, C, and any line m:

$$\alpha \sigma_m = \sigma_{\alpha(m)} \alpha, \qquad\qquad\qquad \alpha \sigma_C = \sigma_{\alpha(C)} \alpha,$$

$$\alpha \tau_{A,B} = \tau_{\alpha(A),\alpha(B)} \alpha, \qquad\qquad \alpha \rho_{C,\Theta} = \rho_{\alpha(C),\Theta} \alpha.$$

   (g) $\rho_{B,\Phi} \rho_{A,-\Phi}$ is the translation that takes A to $\rho_{B,\Phi}(A)$.
   (h) Exactly n of the elements of D_n are involutions.
   (i) Group D_n is a cyclic group with $2n$ elements.
   (j) Group D_n contains exactly n reflections.

7.8. Give the equations for each transformation in D_4.

7.9. In Figure 7.9 sketch a construction for the point Z that is fixed by $\sigma_b \sigma_P \sigma_a$.

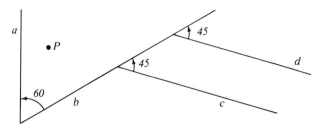

Figure 7.9

7.10. Find Q and Θ such that $\sigma_d \sigma_c \sigma_b \sigma_a = \rho_{Q,\Theta}$ in Figure 7.9.

7.11. In a figure consisting of two points A and B, sketch a construction for point P such that $\tau_{P,B} \rho_{B,60}$ fixes A.

7.12. In a figure consisting of two points A and B off line p, sketch a construction for the line z that is fixed by $\sigma_B \sigma_p \sigma_A$.

7.13. Prove or disprove: Given $\tau_{A,B}$ and nonidentity rotation $\rho_{C,\Theta}$, there is a rotation $\rho_{D,\Phi}$ such that $\tau_{A,B} = \rho_{D,\Phi} \rho_{C,\Theta}$.

7.14. Show that if ρ_1, ρ_2, $\rho_2 \rho_1$, and $\rho_2^{-1} \rho_1$ are rotations, then the centers of ρ_1, $\rho_2 \rho_1$, and $\rho_2^{-1} \rho_1$ are collinear.

7.15. Given four distinct points, find a square such that each of the lines containing a side of the square passes through one of the four given points.

7.16. Describe the symmetry group for each of the eight figures in Figure 7.10.

7.17. Show $\sigma_P \sigma_l \sigma_P \sigma_l \sigma_P \sigma_l \sigma_P \sigma_l \sigma_P$ is a reflection in a line parallel to line l.

Figure 7.10

Chapter 8

Classification of Plane Isometries

§8.1 Glide Reflections

We have classified all the even isometries as translations or rotations. An odd isometry is a reflection or a product of three reflections. Only those odd isometries $\sigma_c \sigma_b \sigma_a$ where a, b, c are neither concurrent nor have a common perpendicular remain to be considered. Although it seems there might be many cases, depending on which of a, b, c intersect or are parallel to which, we shall see this turns out not to be the case. However, we begin with the special case where a and b are perpendicular to c. Then $\sigma_b \sigma_a$ is a translation or *glide* and σ_c is, of course, a reflection. If a and b are distinct lines perpendicular to line c, then $\sigma_c \sigma_b \sigma_a$ is called a ***glide reflection*** with ***axis*** c. We might as well call line m the ***axis*** of σ_m as the reflection and the glide reflection then share the property that the midpoint of any point P and its image under the isometry lies on the axis. To show this holds for the glide reflection, suppose P is any point. See Figure 8.1. Let line l be the perpendicular from P to c. Then there is a line m perpendicular to c such that $\sigma_b \sigma_a = \sigma_m \sigma_l$. If M is the intersection of m and c, then P and M are distinct points such that

$$\sigma_c \sigma_b \sigma_a(P) = \sigma_c \sigma_m \sigma_l(P) = \sigma_c \sigma_m(P) = \sigma_M(P) \neq P.$$

Since $\sigma_c \sigma_b \sigma_a(P) = \sigma_M(P)$ and M is the midpoint of distinct points P and $\sigma_M(P)$, we have shown that glide reflection $\sigma_c \sigma_b \sigma_a$ fixes no point but the midpoint of any point P and its image $\sigma_c \sigma_b \sigma_a(P)$ lies on the axis of the glide reflection. So a glide reflection interchanges the halfplanes of its axis. Hence, any line fixed by the glide reflection must intersect the axis at least twice. That is, the glide reflection can fix no line except its axis. The axis of a glide reflection is the unique line fixed by the glide reflection. We have demonstrated the following properties of a glide reflection.

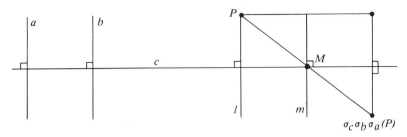

Figure 8.1

Theorem 8.1. *A glide reflection fixes no points. A glide reflection fixes exactly one line, its axis. The midpoint of any point and its image under a glide reflection lies on the axis of the glide reflection.*

If γ is a glide reflection, then there are distinct lines a, b, c such that $\gamma = \sigma_c \sigma_b \sigma_a$ where a and b are perpendicular to c, say at points A and B, respectively. Now, $\sigma_A = \sigma_a \sigma_c = \sigma_c \sigma_a$ and $\sigma_B = \sigma_b \sigma_c = \sigma_c \sigma_b$. Hence,

$$\gamma = \sigma_c(\sigma_b \sigma_a) = (\sigma_c \sigma_b)\sigma_a = \sigma_b(\sigma_c \sigma_a) = (\sigma_b \sigma_a)\sigma_c$$
$$= \sigma_B \sigma_a \qquad = \sigma_b \sigma_A.$$

The first line of these equations tells us that γ is the composite of the glide $\sigma_b \sigma_a$ and the reflection σ_c in either order. More important, the second line tells us that γ is a product $\sigma_B \sigma_a$ with B off a and a product $\sigma_b \sigma_A$ with A off b. See Figure 8.2. We want to show, conversely, that such a product is a glide reflection. Suppose point P is off line l. Let p be the perpendicular from P to l and let m be the perpendicular at P to p. Lines l and m are distinct since P is off l. Furthermore, $\sigma_P \sigma_l = \sigma_p \sigma_m \sigma_l$ and $\sigma_l \sigma_P = \sigma_l \sigma_p \sigma_m = \sigma_p \sigma_l \sigma_m$. Therefore, the products $\sigma_P \sigma_l$ and $\sigma_l \sigma_P$ are glide reflections by the definition of a glide reflection, as desired.

Theorem 8.2. *A glide reflection is the composite of a reflection in some line a followed by a halfturn about some point off a. A glide reflection is the composite of a halfturn about some point A followed by a reflection in some line off A. Conversely, if point P is off line l, then $\sigma_P \sigma_l$ and $\sigma_l \sigma_P$ are glide reflections with axis the perpendicular from P to l.*

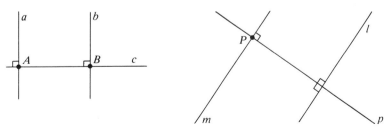

Figure 8.2

Since $\sigma_P\sigma_l$ and $\sigma_l\sigma_P$ are inverses of each other for any point P and line l, the set of all glide reflections has the inverse property. Of course, the set of all glide reflections does not have the closure property because the product of two odd glide reflections must be an even isometry. Since a translation fixing line c commutes with σ_c and since any two translations commute, then a translation fixing line c commutes with any glide reflection with axis c. The square of a glide reflection γ with axis c is just the square of the glide and, hence, a nonidentity translation fixing c. The group $\langle\gamma^2\rangle$, the cyclic group generated by nonidentity translation γ^2, is infinite and is contained in $\langle\gamma\rangle$, the cyclic group generated by γ. So $\langle\gamma^2\rangle$ contains all the even powers of γ, and $\langle\gamma\rangle$ contains all the powers of γ. Since $\langle\gamma^2\rangle$ is infinite, then $\langle\gamma\rangle$ must be infinite. We have proved the following.

Theorem 8.3. *A translation that fixes line c commutes with a glide reflection with axis c. The square of a glide reflection is a nonidentity translation. A glide reflection generates an infinite cyclic group.*

The picture that powers of a glide reflection always bring to mind is Figure 8.3. Each footprint there is the image of the preceding footprint under a glide reflection γ; each left footprint is the image of the preceding left footprint under the translation γ^2.

Figure 8.3

If $\sigma_r\sigma_q\sigma_p$ is a glide reflection, then $\sigma_r\sigma_q\sigma_p$ is not a reflection and the lines p, q, r cannot be either concurrent or parallel. In order to prove the converse, we suppose p, q, r are any lines that are neither concurrent nor have a common perpendicular. We wish to prove that $\sigma_r\sigma_q\sigma_p$ is a glide reflection. First, we consider the case lines p and q intersect at some point Q. Then Q is off r as the lines p, q, r are not concurrent. See Figure 8.4. Let P be the foot of the perpendicular from Q to r, and let m be the line through P and Q. There is a line l through Q such that $\sigma_q\sigma_p = \sigma_m\sigma_l$. Since $p \neq q$, then $l \neq m$ and P is off l. Hence, $\sigma_r\sigma_q\sigma_p = \sigma_r\sigma_m\sigma_l = \sigma_P\sigma_l$ with P off l. Therefore, $\sigma_r\sigma_q\sigma_p$ is a glide reflection by Theorem 8.2.

There remains the case $p \parallel q$. In this case, lines r and q must intersect as otherwise p, q, r have a common perpendicular. Then, by what we just proved in the paragraph above there is some point P off some line l such that $\sigma_p\sigma_q\sigma_r = \sigma_P\sigma_l$. Hence,

$$\sigma_r\sigma_q\sigma_p = (\sigma_p\sigma_q\sigma_r)^{-1} = (\sigma_P\sigma_l)^{-1} = \sigma_l\sigma_P$$

with point P off line l. Therefore, again we have $\sigma_r\sigma_q\sigma_p$ is a glide reflection.

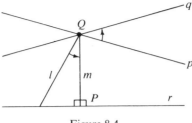

Figure 8.4

Theorem 8.4. *Lines p, q, r are neither concurrent nor have a common perpendicular iff $\sigma_r \sigma_q \sigma_p$ is a glide reflection.*

An immediate corollary of this theorem is that a product of three reflections is a reflection or a glide reflection. Thus, we have a classification of the odd isometries.

Theorem 8.5. *An odd isometry is either a reflection or a glide reflection.*

We finally have ***The Classification Theorem for the Isometries on the Plane***:

Theorem 8.6. *Each nonidentity isometry is exactly one of the following:*

translation,   rotation,   reflection, glide reflection.

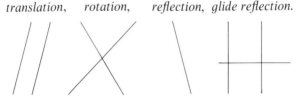

Figure 8.5

Figure 8.5 should not look like mere doodling to you. You should be reminded, in turn, of a translation, a rotation, a reflection, and a glide reflection. Conversely, each of the words *translation, rotation, reflection,* and *glide reflection* should cause a part of the figure to come to mind. As a special case whenever anyone says "halfturn," you should first think "perpendicular lines."

Suppose γ is a glide reflection with axis c and α is an isometry. So $\gamma^2 \neq \iota$. Since $\alpha\gamma\alpha^{-1}$ is an odd isometry that fixes line $\alpha(c)$ but is not an involution, then $\alpha\gamma\alpha^{-1}$ has to be a glide reflection with axis $\alpha(c)$.

Theorem 8.7. *If γ is a glide reflection with axis c and α is an isometry, then $\alpha\gamma\alpha^{-1}$ is a glide reflection with axis $\alpha(c)$.*

At the end of the next section we are going to list all the isometries that take a given point P to a different given point Q. Since this is an excellent review exercise, you may wish to test your skill by doing the exercise before you read the solution.

§8.2 Leonardo's Theorem

Do yourself a favor. Obtain from your library or bookstore a copy of *Symmetry* by the renowned mathematician Hermann Weyl (1885–1955). Read this delightful, popular book published by Princeton University Press in 1952 and still in print. In *Symmetry*, Weyl points out that Leonardo da Vinci (1452–1519) systematically determined the possible symmetries of a central building and how to attach chapels and niches without destroying the symmetry of the nucleus. We shall prove Leonardo's result, which in our terms states that the only finite groups of symmetries are the cyclic groups C_n and the dihedral groups D_n.

Suppose $\mathscr{G}$ is a finite group of isometries. Then $\mathscr{G}$ cannot contain a non-identity translation or a glide reflection, as either of these would generate an infinite subgroup of $\mathscr{G}$. So $\mathscr{G}$ contains only rotations and reflections. We shall consider the case $\mathscr{G}$ contains only rotations and the case $\mathscr{G}$ contains at least one reflection separately.

Suppose $\mathscr{G}$ is a finite group of symmetries that contains only rotations. One possibility is that $\mathscr{G}$ is the identity group C_1. Otherwise, we suppose $\mathscr{G}$ contains a nonidentity rotation $\rho_{A,\Theta}$. Assume $\rho_{B,\Phi}$ is a nonidentity rotation in $\mathscr{G}$ such that $B \neq A$. Then $\mathscr{G}$ must contain the composite

$$\rho_{B,\Phi}^{-1}\rho_{A,\Theta}^{-1}\rho_{B,\Phi}\rho_{A,\Theta},$$

which is a translation by the Angle-addition Theorem (Theorem 7.9) that is not the identity (Theorem 7.7). Since this is impossible, then we must have $B = A$ and all the nonidentity rotations in $\mathscr{G}$ have center A. We note that $\rho_{A,-\Phi}$ is in $\mathscr{G}$ iff $\rho_{A,\Phi}$ is in $\mathscr{G}$ and that all the elements in $\mathscr{G}$ can be written in the form $\rho_{A,\Phi}$ where $0 \leq \Phi < 360$. Let $\rho = \rho_{A,\Phi}$ where Φ has the minimum positive value. If $\rho_{A,\Psi}$ is in $\mathscr{G}$ with $\Psi > 0$, then $\Psi - k\Phi$ cannot be positive and less than Φ for any integer k by the minimality of Φ. So $\Psi = k\Phi$ for some integer k and $\rho_{A,\Psi} = \rho^k$. In other words, the elements of $\mathscr{G}$ are precisely the powers of ρ. We conclude that a finite group of isometries that contains no reflections is a cyclic group C_n for some integer n.

We now turn to the case where $\mathscr{G}$ is a finite group of isometries that contains at least one reflection. Since ι is an even isometry, since an isometry and its inverse have the same parity, and since the product of two even isometries is an even isometry, it follows that the subset of all even isometries in $\mathscr{G}$ forms a finite subgroup of $\mathscr{G}$. By the preceding paragraph, we see that this subgroup must be the cyclic group C_n generated by some rotation ρ, say with center A. So the even isometries in $\mathscr{G}$ are the n rotations $\rho, \rho^2, \ldots, \rho^n$ with $\rho^n = \iota$. Suppose $\mathscr{G}$ has m reflections. If σ is a reflection in $\mathscr{G}$, then the n odd isometries $\rho\sigma$, $\rho^2\sigma$, ..., $\rho^n\sigma$ are in $\mathcal{G}$. So $n \leq m$. However, the m odd isometries multiplied on the right by σ give m distinct even isometries. So $m \leq n$. Hence, $m = n$ and $\mathcal{G}$ contains the $2n$ elements generated by rotation ρ and reflection σ. If $n = 1$, then $\mathcal{G} = \langle\sigma\rangle$. If $n > 1$, then $\rho\sigma$ must be a reflection in a line through the center of A. We conclude that finite group of isometries that contains a reflection is a dihedral group D_n for some integer n.

Compiling the two results above, we have a proof of **Leonardo's Theorem**:

Theorem 8.8. *A finite group of isometries is either a cyclic group C_n or a dihedral group D_n.*

In formulating the definition of the dihedral groups in the preceding chapter, we noticed that a polygon with m vertices has at most $2m$ symmetries. Since the symmetry group of a polygon must then be a finite group, Leonardo's Theorem has the following immediate corollary.

Theorem 8.9. *The symmetry group for a polygon is either a cyclic group or a dihedral group.*

Our set task is to list, without repetition, all the isometries that take a given point P to a different given point Q. To do this we let M be the midpoint of $\overline{PQ}$ and let m be the perpendicular bisector of $\overline{PQ}$. We shall see below that the list consists of exactly the isometries $\sigma_q\sigma_m$ and $\sigma_q\sigma_M$ where q ranges over the set of all the lines that pass through Q. The even isometries and the odd isometries will be considered separately.

Translation $\tau_{P,Q}$ is the unique translation that takes P to Q, and $\tau_{P,Q} = \sigma_q\sigma_m$ where q is the line through Q that is parallel to m. If rotation $\rho_{C,\Theta}$ takes P to Q, then C must be on m, the perpendicular bisector of $\overline{PQ}$, since $CP = CQ$. Since m passes through C, then there is a unique line q through C such that $\rho_{C,\Theta} = \sigma_q\sigma_m$. Then, since $Q = \sigma_q\sigma_m(P) = \sigma_q(Q)$, point Q lies on q. Hence, every rotation taking P to Q has the form $\sigma_q\sigma_m$ where q is a line through Q that intersects m. Conversely, if q is a line through Q that intersects m, then $\sigma_q\sigma_m$ is a rotation taking P to Q. Therefore, the even isometries that take P to Q are exactly the isometries $\sigma_q\sigma_m$ with q on Q.

Reflection σ_m is the unique reflection taking P to Q since m is the unique perpendicular bisector of $\overline{PQ}$. If glide reflection γ with axis c takes P to Q, then M, the midpoint of P and Q, is on c. Since M is on c, there is a unique line q off M such that $\gamma = \sigma_q\sigma_M$. Then, since $Q = \gamma(P) = \sigma_q\sigma_M(P) = \sigma_q(Q)$, point Q lies on q. Hence, every glide reflection taking P to Q has the form $\sigma_q\sigma_M$ where q is a line through Q but off M. Conversely, if q is a line through Q but off M, then $\sigma_q\sigma_M$ is a glide reflection taking P to Q. Now, $\sigma_m = \sigma_q\sigma_M$ with $q = \overleftrightarrow{PQ}$. Therefore, the odd isometries that take P to Q are exactly the isometries $\sigma_q\sigma_M$ with q on Q.

A second, shorter argument than that in the preceding paragraph runs as follows. Suppose α is an odd isometry that takes P to Q. Let $n = \overleftrightarrow{PQ}$. Then $\alpha\sigma_n$ is an even isometry that takes P to Q. By the argument for the even isometries, $\alpha\sigma_n = \sigma_q\sigma_m$ for some line q on Q. So $\alpha = \sigma_q\sigma_m\sigma_n = \sigma_q\sigma_M$ with line q on Q. Conversely, $\sigma_q\sigma_M$ with q on Q is an odd isometry taking P to Q. As above, the odd isometries that take P to Q are exactly the isometries $\sigma_q\sigma_M$ with q on Q.

§8.3 Exercises

8.1. Suppose $\overline{PQ}$ has midpoint M and perpendicular bisector m. Show the set of isometries that take P to Q consists of exactly the isometries $\sigma_m\sigma_p$ and $\sigma_M\sigma_p$ where p ranges over the set of all lines that pass through P.

8.2. If line b intersects lines a and c only at distinct points C and A, respectively, show that the axis of the glide reflection $\sigma_c\sigma_b\sigma_a$ contains the feet of the perpendiculars to a and c from A and C, respectively. This gives an easy construction for the axis of the glide reflection. In Figure 8.6, sketch the axis of $\sigma_c\sigma_b\sigma_a$, the axis of $\sigma_a\sigma_b\sigma_c$, and the axis if $\sigma_a\sigma_c\sigma_b$.

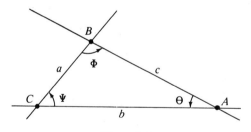

Figure 8.6

8.3. With the notation as in Figure 8.6, show that $\rho_{C,\,2\Psi}\rho_{B,\,2\Phi}\rho_{A,\,2\Theta}$ is a nonidentity translation although $\rho_{A,\,2\Theta}\rho_{B,\,2\Phi}\rho_{C,\,2\Psi} = \iota$.

8.4. Show that translation τ commutes with σ_c iff τ fixes c. Also, that τ commutes with a glide reflection with axis c iff τ fixes c.

8.5. Prove *Hjelmslev's Theorem*: If α is any isometry and l is any line, then there is a line m such that for every point P on l the midpoint of P and $\alpha(P)$ is on m.

8.6. Name the symmetry group for each of the figures in Figure 8.7.

8.7. True or False
    (a) Every isometry is a product of two involutions.
    (b) If $n > 2$, then D_n is generated by two reflections.
    (c) An isometry that does not fix a point is a glide reflection.
    (d) If γ is a glide reflection, then $\gamma(P) \neq P$ for every point P but the midpoint of P and $\gamma(P)$ is on the only line fixed by γ.

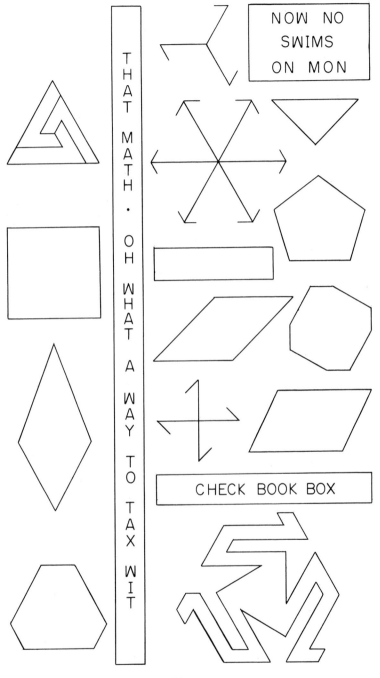

Figure 8.7

(e) If $\gamma = \sigma_l \sigma_P$, then γ is a glide reflection with axis the line through P that is perpendicular to l.

(f) If γ is a glide reflection with axis c and P is a point on c, then there are unique lines l and m such that $\gamma = \sigma_m \sigma_P = \sigma_P \sigma_l$.

(g) If $\sigma_c \sigma_B \sigma_a$ fixes line m, then $\sigma_c \sigma_B \sigma_a$ is a glide reflection with axis m.

(h) If $\sigma_C \sigma_b \sigma_A$ fixes line m, then $\sigma_C \sigma_b \sigma_A$ is a glide reflection with axis m.

(i) A group of isometries that has order 35 must be a cyclic group C_{35}.

(j) Every finite group of isometries is the symmetry group of some polygon.

8.8. Name the finite groups of isometries such that every element in the group fixes given line l. Name the finite groups of isometries such that every element in the group fixes given point P.

8.9. Prove or disprove: If point M is on the axis of glide reflection γ, then there is a point P such that M is the midpoint of P and $\gamma(P)$.

8.10. Prove or disprove: Every glide reflection is a product of three reflections in the three lines containing the sides of some triangle.

8.11. Prove or disprove: If α is an odd isometry, then $\alpha = \sigma_c \sigma_b \sigma_a$ where any one of a, b, c may be arbitrarily chosen.

8.12. Which isometries are dilatations?

8.13. Prove: If τ is a translation, then there is a glide reflection γ such that $\tau = \gamma^2$.

8.14. Prove: If $\gamma^m = \gamma^n$ for glide reflection γ, then $m = n$.

8.15. List all symmetry groups that are the symmetry groups of quadrilaterals and for each group sketch a quadrilateral having that symmetry group.

8.16. If an n-gon has symmetry group C_4, what can be said about n?

8.17. Prove *Lagrange's Theorem*: Let $\mathscr{G}$ be a finite group. Then the order of any subgroup of $\mathscr{G}$ divides the order of $\mathscr{G}$, and the order of any element of $\mathscr{G}$ divides the order of $\mathscr{G}$.

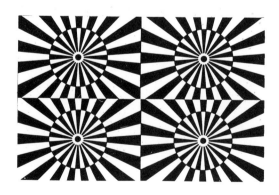

Chapter 9

Equations for Isometries

§9.1 Equations

The equations for a general translation were incorporated in the definition of a translation. Equations for a reflection were determined in Theorem 4.2. We now turn to rotations. Equations for the rotation about the origin through a directed angle of $\Theta°$ are considered first. Let $\rho_{(0,0),\Theta} = \sigma_m\sigma_l$ where l is the X-axis. Then one directed angle from l to m has directed measure $\Theta/2$. From the definition of the trigonometric functions, we know $(\cos(\Theta/2), \sin(\Theta/2))$ is a point on m. So line m has equation $(\sin(\Theta/2))X - (\cos(\Theta/2))Y + 0 = 0$. Hence σ_m has equations

$$x' = x - \frac{2(\sin(\Theta/2))[(\sin(\Theta/2))x - (\cos(\Theta/2))y]}{\sin^2(\Theta/2) + \cos^2(\Theta/2)}$$

$$= [1 - 2\sin^2(\Theta/2)]x + [2\sin(\Theta/2)\cos(\Theta/2)]y$$

$$= (\cos\Theta)x + (\sin\Theta)y,$$

$$y' = y + \frac{2(\cos(\Theta/2))[(\sin(\Theta/2)x - (\cos(\Theta/2))y]}{\sin^2(\Theta/2) + \cos^2(\Theta/2)} = (\sin\Theta)x - (\cos\Theta)y.$$

Since σ_l has equations $x' = x$ and $y' = -y$, then the rotation $\sigma_m\sigma_l$ has equations given by the following theorem.

Theorem 9.1. *In the Cartesian plane, rotation $\rho_{O,\Theta}$ about the origin has equations*

$$\begin{cases} x' = (\cos\Theta)x - (\sin\Theta)y, \\ y' = (\sin\Theta)x + (\cos\Theta)y. \end{cases}$$

71

Since $\rho_{(h, k), \Theta} = \tau_{(0, 0), (h, k)} \rho_{(0, 0), \Theta} \tau_{(h, k), (0, 0)}$ by Theorem 7.6, the equations for rotation $\rho_{(h, k), \Theta}$ about the point (h, k) are easily obtained by composing three sets of equations. The rotation has equations

$$\begin{cases} x' = (\cos \Theta)(x - h) - (\sin \Theta)(y - k) + h, \\ y' = (\sin \Theta)(x - h) + (\cos \Theta)(y - k) + k. \end{cases}$$

It is easier to derive these equations from Theorem 9.1 when needed than to remember them as a special theorem. These equations for the rotation $\rho_{(h, k), \Theta}$ have the form

$$\begin{cases} x' = (\cos \Theta)x - (\sin \Theta)y + r, \\ y' = (\sin \Theta)x + (\cos \Theta)y + s, \end{cases}$$

which, conversely, are the equations of a rotation unless $\Theta° = 0°$, since given r, s, and Θ there are unique solutions for h and k given by

$$r = h(1 - \cos \Theta) + k(\sin \Theta) \quad \text{and} \quad s = h(-\sin \Theta) + k(1 - \cos \Theta)$$

unless $\Theta° = 0°$. In case $\Theta° = 0°$, the equations above are those of a general translation. Note that Theorem 7.9, the Angle-addition Theorem, can thus be proved directly by composing two sets of equations of this last form. Since the even isometries are the translations and the rotations, setting $a = \cos \Theta$ and $b = \sin \Theta$ we have the following theorem.

Theorem 9.2. *The general equations for an even isometry on the Cartesian plane are*

$$\begin{cases} x' = ax - by + c, \\ y' = bx + ay + d, \end{cases} \quad \text{with } a^2 + b^2 = 1$$

and, conversely, such equations are those of an even isometry.

If α is an odd isometry and l any line, then α is the product of even isometry $\sigma_l \alpha$ followed by σ_l. Taking l as the X-axis, we have any odd isometry is the product of an even isometry followed by the reflection in the X-axis. This observation, together with Theorem 9.2, gives the following theorem where the positive sign applies when the isometry is even and the negative sign applies when the isometry is odd.

Theorem 9.3. *The general equations for an isometry on the Cartesian plane are*

$$\begin{cases} x' = ax - by + c, \\ y' = \pm(bx + ay + d), \end{cases} \quad \text{with } a^2 + b^2 = 1$$

and, conversely, such equations are those of an isometry.

It is certainly easy to distinguish between equations for translations and equations for rotations. Criteria for distinguishing between equations for reflections and equations for glide reflections are left for the exercises in Section 9.3.

§9.2 Supplementary Exercises (Chapters 1–8)

(a) What is the next figure in the sequence illustrated in Figure 9.1?

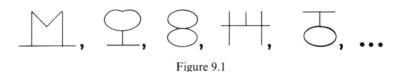

Figure 9.1

(b) In Figure 9.2, sketch and find the length of the shortest path from B to C to D to E where points C and D are 4 units apart on line r.

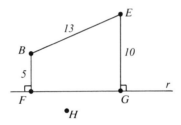

Figure 9.2

(c) In Figure 9.2, sketch a pentagon having points E, B, F, H, G as midpoints of the sides taken in order.

(d) In Figure 9.2, if $\sigma_G \sigma_P \sigma_E(B) = F$, then sketch a construction for point P.

(e) Sketch a construction of a path of a hole-in-one in the miniature golf course in Figure 9.3.

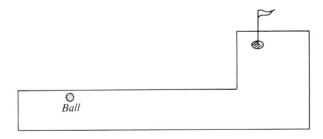

Figure 9.3

(f) List as many infinite groups of isometries as you can think of in ten minutes.

(g) The equation $S = [(a + b - c)(a - b + c)/a]^{1/2}$ cannot be a general formula for the area S of triangles having sides of length a, b, c. How can you tell this at a glance?

(h) Prove or disprove: If s is the set of all images of point A under the elements of some nonidentity group $\mathscr{G}$ of isometries, then $\mathscr{G}$ may be a proper subgroup of the group of symmetries of s.

(i) If $A = (1, 2)$ and $B = (-3, 6)$, find C such that $\sigma_B \sigma_A(C) = (4, -2)$. If $E = (2, 3)$, then find D such that $\sigma_E \sigma_D((1, -2)) = (3, 5)$.

(j) If $A = (3, 0)$, find equations for lines m and n such that $(2, -4)$ is on m and $\sigma_A = \sigma_n \sigma_m$.

(k) If point A is off each of the intersecting lines p and q, find all pairs P and Q such that P is on p, Q is on q, and A is the midpoint of P and Q.

(l) Sketch the shortest path from B to E that crosses each of the three indicated "rivers" in Figure 9.4 at right angles.

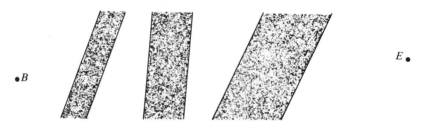

Figure 9.4

(m) Suppose α and β are two symmetries for set s of points and $\alpha(P) = \beta(P)$ for every point P in s. What can you say about s?

(n) Prove or disprove: If C_n is a group of symmetries of an n-gon, then D_n is the symmetry group for the n-gon.

(o) Prove or disprove: If C_m or D_m is the symmetry group for an n-gon, then $n = km$ for some integer k.

(p) Prove or disprove: If $\rho_{B,\Phi}\rho_{A,\Theta} = \rho_{C,\Theta+\Phi}$, $\rho_{A,\Theta}\rho_{B,\Phi} = \rho_{D,\Theta+\Phi}$, and $m = \overleftrightarrow{AB}$, then $D = \sigma_m(C)$.

(q) Given point M, a line, and a circle, find points L and C such that L is on the line, C is on the circle, and M is the midpoint of L and C.

(r) Given a line l and two circles, find squares $\square PQRS$ such that P and R are on the line, Q is on one circle, and S is on the other circle.

(s) The pool table in Figure 9.5 has pockets only at the corners. Into which pocket does the ball go?

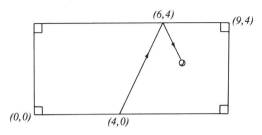

Figure 9.5

(t) *Thomsen's relation:* Prove that for any lines a, b, c:

$$\sigma_c\sigma_a\sigma_b\sigma_c\sigma_a\sigma_b\sigma_a\sigma_b\sigma_c\sigma_a\sigma_b\sigma_c\sigma_b\sigma_a\sigma_c\sigma_b\sigma_a\sigma_b\sigma_a\sigma_c\sigma_b\sigma_a = \iota.$$

(u) Name the symmetry group for each of the nine figures in Figure 9.6. These are taken from Hajime Ōuchi's *Japanese Optical and Geometrical Art* in the Dover Pictorial Archive Series.

Figure 9.6

(v) Given a point, a line, and a circle, find the squares having one vertex the point, one vertex on the line, and a third vertex on the circle.

(w) Given $\triangle ABC$, if $\triangle BCD$ and $\triangle ACE$ are equilateral with A and D on opposite sides of $\overleftrightarrow{BC}$ and with B and E on opposite sides of $\overleftrightarrow{AC}$, then show $AD = BE$.

(x) If point P is on side $\overline{BC}$ of acute triangle $\triangle ABC$, find point Q on $\overline{AC}$ and point R on $\overline{AB}$ such that $\triangle PQR$ has minimal perimeter.

(y) Given acute triangle $\triangle ABC$, find points P, Q, R with P on $\overline{BC}$, Q on $\overline{AC}$, and R on $\overline{AB}$ such that $\triangle PQR$ has minimal perimeter (*Fagnano's problem*).

(z) A bee and a lump of sugar are inside an acute triangle. The bee wishes to reach the sugar but must touch all three sides of the triangle before coming to the sugar. What is the shortest path for the bee?

§9.3 Exercises

9.1. What are the equations for each of $\sigma_{O,90}$, $\rho_{O,180}$, and $\rho_{O,270}$?

9.2. What are the equations for $\rho_{O,60}$ and for $\rho_{(1,-2),60}$?

9.3. Explain the test for symmetry of a curve with respect to the X-axis, namely that the equation of the curve remains unchanged when "Y" is replaced by "$-Y$" throughout the equation. What are the tests for symmetry with respect to each of the Y-axis, the origin, the line with equation $Y = X$, and the line with equation $Y = -X$?

9.4. Show $x' = ax + by + c$ and $y' = bx - ay + d$ are general equations for an odd isometry when $a^2 + b^2 = 1$.

9.5. If $x' = ax + by + c$ and $y' = bx - ay + d$ with $a^2 + b^2 = 1$ are the equations for isometry α, show α is a reflection iff $ac + bd + c = 0$ and $ad - bc - d = 0$.

9.6. True or False
(a) $x' = -x + 6$ and $y' = -y - 7$ are equations for a rotation.
(b) $x' = px - qy + r$ and $y' = qx + py + s$ are equations for an even isometry.
(c) $x' = -px - qy - r$ and $y' = qx - py - s$ are equations for an even isometry if $p^2 + q^2 = 1$.
(d) $x' = -ax + by + c$ and $y' = bx + ay + d$ are equations for an odd isometry if $a^2 + b^2 = 1$.
(e) $x' = \pm ax - by + c$ and $y' = \pm bx + ay + d$ are equations for an isometry if $a^2 + b^2 = 1$.
(f) $x' = x \cos 2\Phi + y \sin 2\Phi$ and $y' = x \sin 2\Phi - y \cos 2\Phi$ are equations of the reflection in the line m through the origin such that a directed angle from the X-axis to m is $\Phi°$.
(g) If m is any line, then every odd isometry is the product of σ_m followed by an even isometry.
(h) If m is any line, then every odd isometry is the product of an even isometry followed by σ_m.
(i) If τ is a translation, then $\tau\sigma_P$ is the halfturn about the midpoint of P and $\tau(P)$.
(j) If τ is a nonidentity translation that fixes line c and line l is perpendicular to c at point P, then $\tau\sigma_l$ is the reflection in the perpendicular bisector of P and $\tau(P)$.

9.7. If $x' = 3x/5 + 4y/5$ and $y' = 4x/5 - 3y/5$ are equations for σ_m, then find m.

9.8. If $x' = -3x/5 + 4y/5$ and $y' = -4x/5 - 3y/5$ are equations for isometry α, then describe α where $(3/5, 4/5) = (\cos \Phi, \sin \Phi)$.

9.9.   If $2x' = -3^{1/2}x - y + 2$ and $2y' = x - 3^{1/2}y - 1$ are equations for $\rho_{P,\Theta}$, then find P and Θ.

9.10. If $x' = x \cos \Theta - y \sin \Theta + r$ and $y' = x \sin \Theta + y \cos \Theta + s$ are equations for nonidentity rotation $\rho_{P,\Theta}$, then find P.

9.11. If $x' = ax + by + c$ and $y' = bx - ay + d$ are equations for σ_m, then find m.

9.12. If $x' = ax + by + c$ and $y' = bx - ay + d$ are equations for glide reflection γ, show γ has axis with equation

$$2bX - 2(a + 1)Y + (ad - bc + d) = 0$$

unless $b = 0$ and $a = -1$, in which case the axis has equation $X = c/2$.

9.13. If $x' = a_i x + b_i y + c_i$ and $y' = d_i x + e_i y + f_i$ are equations for transformation α_i and $\alpha_3 = \alpha_2\alpha_1$, then show

$$\begin{bmatrix} a_3 & b_3 & c_3 \\ d_3 & e_3 & f_3 \\ 0 & 0 & 1 \end{bmatrix} = \begin{bmatrix} a_2 & b_2 & c_2 \\ d_2 & e_2 & f_2 \\ 0 & 0 & 1 \end{bmatrix}\begin{bmatrix} a_1 & b_1 & c_1 \\ d_1 & e_1 & f_1 \\ 0 & 0 & 1 \end{bmatrix}.$$

9.14. Prove Theorem 7.9, the Angle-addition Theorem, directly from equations for an even isometry.

9.15. With $x = r \cos \Phi$, $y = r \sin \Phi$, $x' = r(\cos \Phi + \Theta)$, and $y' = r(\sin \Phi + \Theta)$, use polar coordinates to prove Theorem 9.1.

9.16. Each of the patterns indicated in Figure 9.7 is assumed to extend infinitely far to the right and to the left. For each of the four patterns, assume a convenient coordinate system and then give the equations for all the symmetries of the indicated pattern.

Figure 9.7

Chapter 10

The Seven Frieze Groups

§10.1 Frieze Groups

Around the frieze of an older building there is often a pattern formed by the repetition of some figure or motif over and over again. The essential property of an ornamental frieze pattern is that it is left fixed by some "smallest translation." We can call AB the **length** of translation $\tau_{A,B}$ and say $\tau_{A,B}$ is **shorter than** $\tau_{C,D}$ if $AB < CD$. Other symmetries in addition to translations are often apparent in a frieze as well. Of course, there is infinite variety in the subject for such patterns. However, by discounting the scale and subject matter and by considering only the symmetries under which such patterns are left invariant, we shall see that there are only seven possible types of ornamental frieze patterns.

If isometries α and σ_P are in group $\mathcal{G}$ of isometries, then $\sigma_{\alpha(P)}$ is also in $\mathcal{G}$ because the product $\alpha\sigma_P\alpha^{-1}$ must be in $\mathcal{G}$. Similarly, if isometries α and σ_l are in group $\mathcal{G}$ of isometries, then $\sigma_{\alpha(l)}$ is also in $\mathcal{G}$ because the product $\alpha\sigma_l\alpha^{-1}$ must be in $\mathcal{G}$. Since the symmetries of any set of points form a group, we have proved the following.

Theorem 10.1. *If P is a point of symmetry for set s of points and α is a symmetry of s, then $\alpha(P)$ is a point of symmetry for $\mathcal{S}$. If l is a line of symmetry for set s of points and α is a symmetry of s, then $\alpha(l)$ is a line of symmetry for s.*

A group of isometries that fix a given line c and whose translations form an infinite cyclic group is a **frieze group** with **center** c. Let τ be a nonidentity translation that fixes a given line c. We shall determine all frieze groups $\mathcal{F}$ with center c and whose translations form the infinite cyclic group generated by τ. There will be seven of them. For each group, we shall give a frieze pattern having that group as its group of symmetries. We shall also state in italics criteria which distinguish those patterns having that particular symmetry group.

The following notation will be used throughout the development. We begin by choosing a point A on line c as follows. If $\mathcal{F}$ contains halfturns, then A is chosen to be the center of one of these halfturns; if $\mathcal{F}$ contains no half-turns but does contain reflections in lines perpendicular to c, then A is chosen to be the intersection of one of these lines and c; otherwise, point A is chosen to be any point on c. Let $A_i = \tau^i(A)$. So $A_0 = A$. Since $\tau^n(A_i) = \tau^{n+i}(A)$, then every translation in $\mathcal{F}$ must take each A_i to some A_j. Let M be the midpoint of A and A_1, and let $M_i = \tau^i(M)$. So M_i is the midpoint of A_i and A_{i+1} and also the midpoint of A_0 and A_{2i+1}.

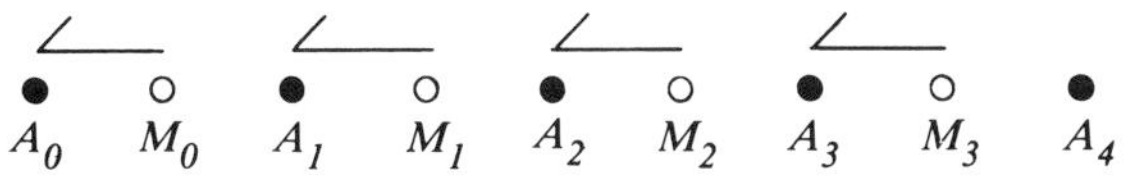

Figure 10.1

One possibility for $\mathcal{F}$ is just the group generated by τ. Let $\mathcal{F}_1 = \langle \tau \rangle$. *A frieze pattern having $\mathcal{F}_1$ as its symmetry group has no point of symmetry, has no line of symmetry, and is not fixed by a glide reflection.* See Figure 10.1, where here, as in the next six figures, the solid dots A_i and the open dots M_i are not to be considered part of the pattern illustrated.

Other than translations, the only even isometries that fix center c are the halfturns with center on c. Suppose $\mathcal{F}$ contains a halfturn. Then σ_A is in $\mathcal{F}$ by the choice of A. Also, σ_M is in $\mathcal{F}$ as σ_M is the product $\tau\sigma_A$. By the first theorem, then $\mathcal{F}$ contains the halfturn about each A_i and about each M_i. Now suppose P is the center of some halfturn in $\mathcal{F}$. Then the translation $\sigma_P\sigma_A$ is in $\mathcal{F}$. So $\sigma_P\sigma_A(A) = A_n$ for some n. Then $\sigma_P(A) = A_n$, and P is the midpoint of A and A_n. Hence $\mathcal{F}$ contains exactly those halfturns that have center A_i and those that have center M_i. Let $\mathcal{F}_2 = \langle \tau, \sigma_A \rangle$. Since $\tau\sigma_A$ is an involution, then $\tau\sigma_A = \sigma_A\tau^{-1}$. So every element in $\mathcal{F}_2$ is of the form τ^i or $\sigma_A\tau^i$. Every element in $\mathcal{F}_2$ is of the form $\sigma_A^j\tau^i$. Also, $\mathcal{F}_2 = \langle \sigma_A, \sigma_M \rangle$ since $\sigma_M\tau = \sigma_A$. *A frieze pattern having $\mathcal{F}_2$ as its group of symmetries has a point of symmetry but no line of symmetry.* See Figure 10.2.

Figure 10.2

If $\mathcal{F}$ contains only even isometries, then $\mathcal{F}$ must be one of $\mathcal{F}_1$ or $\mathcal{F}_2$. The other possibilities for $\mathcal{F}$ are obtained by augmenting $\mathcal{F}_1$ or $\mathcal{F}_2$ with odd isometries. We first consider adding reflections. Recall that σ_l fixes c iff $l = c$ or $l \perp c$. Let $\mathcal{F}_1^1 = \langle \tau, \sigma_c \rangle$. Since $\tau\sigma_c = \sigma_c\tau$, then $\mathcal{F}_1^1$ is abelian and every element is of the form $\sigma_c^j\tau^i$. If $n \neq 0$, then $\mathcal{F}_1^1$ contains the glide

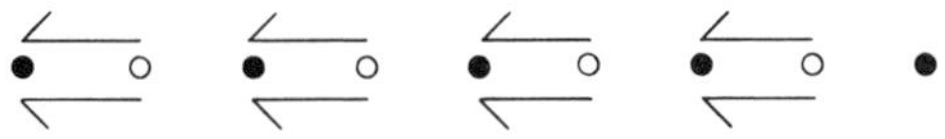

Figure 10.3

reflection with axis c that takes A to A_n. *A frieze pattern having $\mathcal{F}_1^1$ as its symmetry group has no point of symmetry and the center is a line of symmetry.* See Figure 10.3.

Let $\mathcal{F}_2^1 = \langle \tau, \sigma_A, \sigma_c \rangle$. Since σ_c commutes with both τ and σ_A, then every element of $\mathcal{F}_2^1$ is of the form $\sigma_c^k \sigma_A^j \tau^i$. If $n \neq 0$, then $\mathcal{F}_2^1$ contains the glide reflection $\sigma_c \tau^n$ with axis c that takes A to A_n. Also, $\mathcal{F}_2^1$ contains $\tau^{2i} \sigma_A \sigma_c$, which is the reflection in the line perpendicular to c at A_i, and $\mathcal{F}_2^1$ contains $\tau^{2i+1} \sigma_A \sigma_c$, which is the reflection in the line perpendicular to c at M_i. If a is the line perpendicular to c at A, then $\mathcal{F}_2^1 = \langle \tau, \sigma_a, \sigma_c \rangle$. *A frieze pattern having $\mathcal{F}_2^1$ as its symmetry group has a point of symmetry and the center is a line of symmetry.* See Figure 10.4.

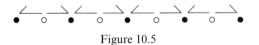

Figure 10.4

Suppose $\mathcal{F}$ does not contain a halfturn but does contain the reflection in a line a that is perpendicular to c. In this case, we suppose A is on a. Then $\mathcal{F}$ contains $\tau^{2i} \sigma_a$, which is the reflection in the line perpendicular to c at A_i, and $\mathcal{F}$ contains $\tau^{2i+1} \sigma_a$, which is the reflection in the line perpendicular to c at M_i. Assume $\mathcal{F}$ contains another reflection σ_l. Then $l \neq c$ as the halfturn $\sigma_c \sigma_a$ is not in $\mathcal{F}$. So $l \perp c$. Then $\mathcal{F}$ contains the translation $\sigma_l \sigma_a$, which must take A to A_n for some n. So $\sigma_l(A) = A_n$ for some n with $n \neq 0$, and l is perpendicular to c at some A_i or at some M_i. Therefore, $\mathcal{F}$ must contain exactly those reflections in lines perpendicular to c at A_i for each i and those reflections in lines perpendicular to c at M_i for each i. We have now considered all possible cases of adding reflections to $\mathcal{F}_1$. Let $\mathcal{F}_1^2 = \langle \tau, \sigma_a \rangle$ where a is perpendicular to c at A. Since $\tau \sigma_a = \sigma_a \tau^{-1}$, then every element of $\mathcal{F}_1^2$ is of the form $\sigma_a^j \tau^i$. $\mathcal{F}_1^2$ does not contain σ_c but does contain the reflections in the lines that are perpendicular to c at A_i or M_i. *A frieze pattern having $\mathcal{F}_1^2$ as its symmetry group has no point of symmetry, has a line of symmetry, but the center is not a line of symmetry.* See Figure 10.5.

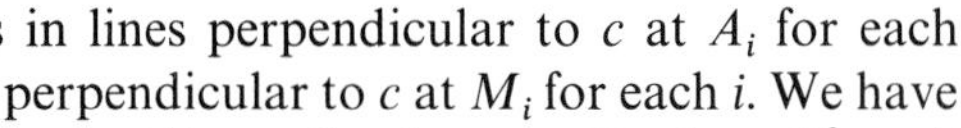

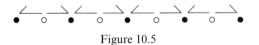

Figure 10.5

Now suppose $\mathcal{F}$ does contain a halfturn and the reflection in a line q. If $q = c$, q is perpendicular to c at A_i, or q is perpendicular to c at M_i, then we are back to $\mathcal{F}_2^1$. To obtain something new, we must suppose q is off each A_i and off each M_i. Since $\sigma_q(A)$ must be the center of a halfturn in $\mathcal{F}$ by the first theorem (with $\alpha = \sigma_q$), the only remaining possibility is that q is the perpendicular bisector of $\overline{AM}_i$ for some i. Because of the halfturns in $\mathscr{F}$, the second part of the first theorem then requires that $\mathscr{F}$ contain the reflection in the perpendicular bisector of $\overline{AM}_i$ for each i. Hence, in particular, $\mathscr{F}$ contains σ_p where p is the perpendicular bisector of $\overline{AM}$. If line a is perpendicular to c at A, then $\mathscr{F}$ cannot contain both σ_p and σ_a as the translation $\sigma_p\sigma_a$ would take A to M, which is impossible. Also, since $\sigma_p\sigma_a = \sigma_p\sigma_c\sigma_A$, then $\mathscr{F}$ cannot contain both σ_p and σ_c. We have now considered all possible cases of adding reflections to $\mathscr{F}_2$. Let $\mathscr{F}_2^2 = \langle \tau, \sigma_A, \sigma_p \rangle$ where p is the perpendicular bisector of $\overline{AM}$. Note that $\mathscr{F}_2^2$ contains the glide reflection $\sigma_p\sigma_A$ with axis c that takes A to M. Let $\gamma = \sigma_p\sigma_A$. Since $\tau = \gamma^2$ and $\sigma_p = \gamma\sigma_A$, then $\mathscr{F}_2^2 = \langle \gamma, \sigma_A \rangle$. $\mathscr{F}_2^2$ does not contain σ_c. *A frieze pattern having $\mathscr{F}_2^2$ as its symmetry group has a point of symmetry, has a line of symmetry, but the center is not a line of symmetry.* See Figure 10.6 but ignore the dots.

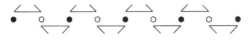

Figure 10.6

We have considered all possibilities for $\mathscr{F}$ that do not necessarily contain a glide reflection. Now suppose $\mathscr{F}$ contains the glide reflection α. Then α has axis c and α^2 is a translation that fixes c. We have two cases: $\alpha^2 = \tau^{2n}$ and $\alpha^2 = \tau^{2n+1}$ for some integer n. Suppose $\alpha^2 = \tau^{2n}$. Since α and τ commute, then $(\alpha\tau^{-n})^2$ is the identity. So the odd involutory isometry $\alpha\tau^{-n}$ must be σ_c. Hence $\alpha = \sigma_c\tau^n$. In this case $\mathscr{F}$ contains σ_c and $\sigma_c\tau^m$ for each integer m. If $\mathscr{F}$ does not contain a halfturn, then we are back to $\mathscr{F}_1^1$; if $\mathscr{F}$ does contain a halfturn, then we are back to $\mathscr{F}_2^1$. Now suppose $\alpha^2 = \tau^{2n+1}$. Then $(\tau^{-n}\alpha)^2$ is τ. Let $\gamma = \tau^{-n}\alpha$. Then γ is an odd isometry whose square is τ. Hence γ must be the unique glide reflection with axis c that takes A to M. Since $\gamma^{2m} = \tau^m$ and $\gamma^{2m+1} = \tau^m\gamma$, the glide reflections in $\mathscr{F}$ are exactly those of the form $\tau^m\gamma$. Let $\mathscr{F}_1^3 = \langle \gamma \rangle$ where γ is the glide reflection with axis c such that $\gamma^2 = \tau$. *A frieze pattern having $\mathscr{F}_1^3$ as its symmetry group has no point of symmetry, has no line of symmetry, but is fixed by a glide reflection.* See Figure 10.7 but ignore the dots.

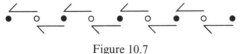

Figure 10.7

Suppose $\mathscr{F}$ contains isometries in addition to those generated by the glide reflection γ with axis c where $\gamma^2 = \tau$. Since the square of the translation $\sigma_c\gamma$ is τ, then $\sigma_c\gamma$ is not in $\langle \tau \rangle$. So σ_c cannot be in $\mathscr{F}$. If $\mathscr{F}$ contains σ_l with $l \perp c$, then $\mathscr{F}$ contains the halfturn $\sigma_l\gamma$. If $\mathscr{F}$ contains a halfturn, then $\mathscr{F}$ must contain σ_A. In this case, $\mathscr{F}$ contains σ_A and the glide reflection γ with center c such that $\gamma^2 = \tau$. Hence $\mathscr{F}$ is $\mathscr{F}_2^2$. We have finally run out of possibilities. Group $\mathscr{F}$ must be one of the seven groups given above, and we have our theorem.

Theorem 10.2. *Let $\mathcal{F}$ be a frieze group with center c whose translations form the group generated by translation τ. If $\mathcal{F}$ contains a halfturn, suppose $\mathcal{F}$ contains σ_A; if $\mathcal{F}$ contains a reflection in a line perpendicular to c, suppose $\mathcal{F}$ contains σ_a with a $\perp$ c. Let γ be the glide reflection with axis c such that $\gamma^2 = \tau$. Then, $\mathcal{F}$ is one of the seven distinct groups defined as follows.*

$$\mathcal{F}_1 = \langle \tau \rangle, \qquad \mathcal{F}_1^1 = \langle \tau, \sigma_c \rangle, \qquad \mathcal{F}_1^2 = \langle \tau, \sigma_a \rangle, \qquad \mathcal{F}_1^3 = \langle \gamma \rangle,$$

$$\mathcal{F}_2 = \langle \tau, \sigma_A \rangle, \qquad \mathcal{F}_2^1 = \langle \tau, \sigma_A, \sigma_c \rangle, \qquad \mathcal{F}_2^2 = \langle \gamma, \sigma_A \rangle.$$

The seven types of ornamental frieze groups are illustrated in Figure 10.8.

$\mathcal{F}_1$ F F F F F F F F $\mathcal{F}_2$ S S S S S S S S

$\mathcal{F}_1^1$ D D D D D D D D $\mathcal{F}_2^1$ I I I I I I I I

$\mathcal{F}_1^2$ A A A A A A A A $\mathcal{F}_2^2$ M W M W M W M W

$\mathcal{F}_1^3$ D W D M D W D M

Figure 10.8

In the next chapter we shall use the fact that the only frieze group containing a glide reflection but no reflections is $\mathcal{F}_1^3$. This follows from the observations that only $\mathcal{F}_1$, $\mathcal{F}_2$, and $\mathcal{F}_1^3$ contain no reflections while only $\mathcal{F}_1$ and $\mathcal{F}_2$ contain no odd isometries. The subgroup relationship among the frieze groups is given in Figure 10.9.

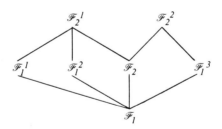

Figure 10.9

§10.2 Frieze Patterns

Can you determine the frieze group that is the symmetry group for each of the seven types of frieze patterns as illustrated in Figure 10.10? To make a key to the frieze patterns, we can ask the following questions.

(1) Is there a point of symmetry?
(2) Is the center a line of symmetry?
(3) Is there any line of symmetry?
(4) Is the pattern fixed by a glide reflection?

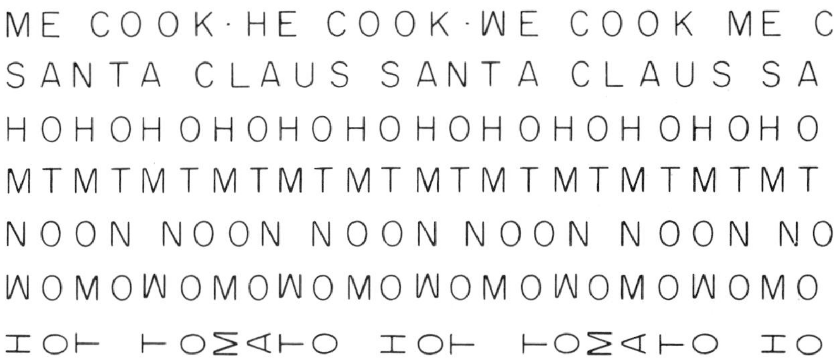

ME COOK·HE COOK·WE COOK ME C

SANTA CLAUS SANTA CLAUS SA

HOH OH OHOHOHO HOH OHOH OHOH O

MTMTMTMTMTMT MTMTMTMTMTMT

NOON NOON NOON NOON NOON NO

WOMOWOMOWOMOWOMOWOMOWOMO

ⅠOⱵ ⱵOƧ∢ⱵO ⅠOⱵ ⱵOƧ∢ⱵO ⅠO

Figure 10.10

Knowing the answers to these questions allows us to determine the symmetry group of the frieze. These questions are based on the criteria in italics from the previous section. The key is given in Figure 10.11. Check the key against Figure 10.8 and then classify the patterns in Figure 10.10. With very little practice you will be able to classify at least four or five patterns in a minute, although at first you will probably need more time than that.

Since there are only seven frieze groups, only three questions are necessary to make a key. However, the four question key seems easier to use. This also agrees with the notation for the groups, which is due to the Hungarian mathematician Fejes Tóth. The subscript 2 or 1 to the $\mathscr{F}$ indicates whether or not there is a halfturn in the group. ·The superscript 1 indicates the center

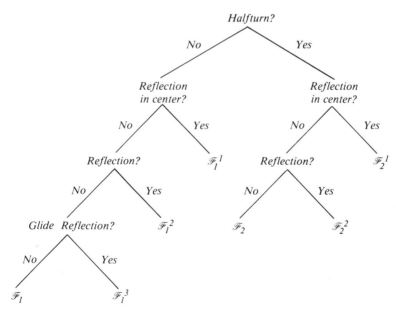

Figure 10.11

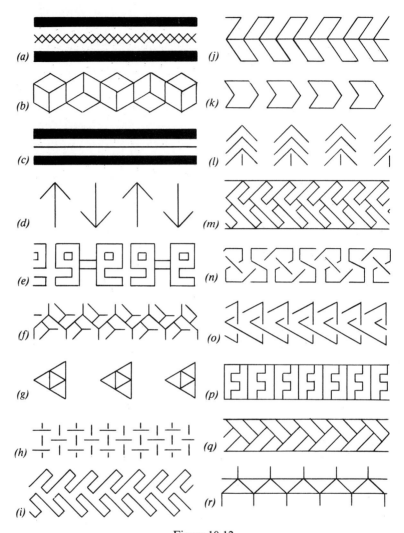

Figure 10.12

is a line of symmetry. The superscript 2 indicates the center is not a line of symmetry but there is a line of symmetry perpendicular to the center. The superscript 3 is left for the special case that the group is generated by a glide reflection.

More patterns for practice in classifying frieze patterns by their symmetry group are given in Figure 10.12 above. However, beware, one of these patterns is not a frieze pattern. Which one? Why? Perhaps the best way to get a feeling for the patterns is to make some yourself. You will be asked to do just this in one of the exercises. Of course, frieze patterns are used in many places other than about the friezes of buildings. Be on the lookout for the frieze patterns you encounter in your surroundings.

§10.3 Exercises

10.1. Name the frieze group for each pattern in Figure 10.10.

10.2. Name the frieze group for each of the frieze patterns in Figure 10.12.

10.3. Make two patterns something like those in Figure 10.12 for each of the seven frieze groups.

10.4. The four patterns in Figure 10.13 use the same motif as those given in Figures 10.1 through 10.6. Name the symmetry group for each of these frieze patterns.

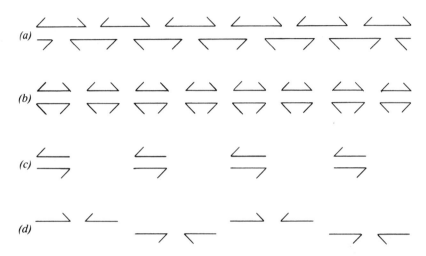

Figure 10.13

10.5. True or False
   (a) Any isometry in a frieze group must fix the set of all centers of the halfturns in the group.
   (b) A glide reflection in a frieze group containing a halfturn must fix the set of all centers of halfturns in the group.
   (c) If γ is a glide reflection such that $\gamma^2 = \tau_{A,B}$, then γ has axis $\overleftrightarrow{AB}$.
   (d) A frieze group containing a glide reflection contains the group $\mathcal{F}_1^1$ or the group $\mathcal{F}_1^3$.
   (e) If a frieze group contains a glide reflection α, then α^2 generates the subgroup of all translations in the frieze group.
   (f) If points A and B are distinct, then there is a unique glide reflection γ with axis $\overleftrightarrow{AB}$ such that $\gamma^2 = \tau_{A,B}$.
   (g) Frieze group $\mathcal{F}_2^2$ contains $\mathcal{F}_2^1$.
   (h) Frieze group $\mathcal{F}_1^2$ contains $\mathcal{F}_1^1$.
   (i) Frieze groups $\mathcal{F}_1$, $\mathcal{F}_2$, and $\mathcal{F}_1^2$ are the only frieze groups that do not contain a glide reflection.
   (j) If α and σ_l are symmetries of a given set of points and α takes line l to line m, then σ_m is in the symmetry group for the set of points.

10.6. Name the frieze group for each of the ten patterns formed by repeating one of the letters from Figure 6.6, e.g.: FFFFF.

10.7. Make a three question key to the frieze patterns.

10.8. Find all possible groups of symmetries that fix a given line but do not contain a translation. For each of these groups, give a figure that could be on the frieze of some building.

10.9. Name the frieze group for each of the following patterns:

(a) AAAAAAAA   (b) F F F F
     AAAAAAAA,   F F F F,
(c) EEEEEEEE   (d) E E E E
     EEEEEEEE,   E E E E,
(e) NNNNNNNN   (f) N N N N
     NNNNNNNN,   N N N N,
(g) DDDDDDDD   (h) D D D D
     DDDDDDDD,   D D D D,
(i) HHHHHHHH   (j) H H H H
     HHHHHHHH,   H H H H.

10.10. Two mathematicians look at a frieze pattern and disagree on the frieze group. Give an example of such a frieze pattern.

10.11. Prove or disprove: A set of points in the interior of a circle cannot have two points of symmetry.

10.12. Figure 10.14 illustrates a pattern with symmetry group $\mathscr{F}_1$. For each of the other six frieze groups, complete Figure 10.14 to a pattern having that group as its symmetry group by adding as little as possible. (There are to be at least four repetitions.)

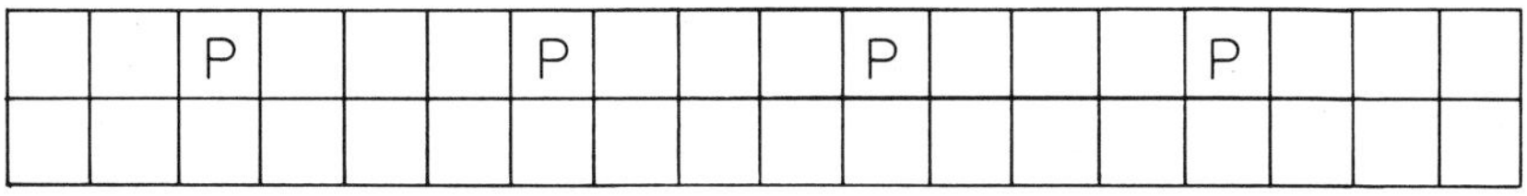

Figure 10.14

10.13. Using a grid like that in Figure 10.14, illustrate each of the seven types of frieze patterns by filling each of the boxes with "X" or "O".

10.14. Prove or disprove: A group generated by two glide reflections with the same axis is a frieze group.

10.15. Find all glide reflections γ such that $\gamma^2 = \tau$ for given nonidentity translation τ.

10.16. Show that each frieze group $\mathscr{F}_2$, $\mathscr{F}_1^2$, and $\mathscr{F}_2^2$ is generated by two involutions. Show $\mathscr{F}_2^1$ is generated by three involutions.

10.17. Prove: If P is the center of a rotation of 90° in a given group of isometries and the group contains an isometry that takes P to Q, then Q is also the center of a rotation of 90° in the group.

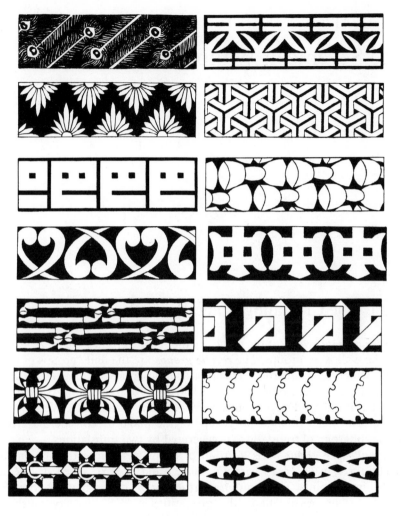

Figure 10.15

10.18. Name the symmetry group associated with each of the fourteen frieze patterns indicated in Figure 10.15. These are taken from Theodore Menten's *Japanese Border Designs* in the Dover Pictorial Archive Series.

Chapter 11

The Seventeen Wallpaper Groups

§11.1 The Crystallographic Restriction

The **ornamental groups** of the plane are the rosette groups, the frieze groups, and the wallpaper groups. The **rosette groups** are the finite groups of isometries, which by Leonardo's Theorem are the groups C_n and D_n. A frieze group is a group of isometries whose subgroup of translations is generated by one translation. Frieze groups were treated in Chapter 10. We now turn to the last of the ornamental groups of the plane by considering groups whose subgroup of translations is generated by two translations.

A **wallpaper group** $\mathcal{W}$ is a group of isometries whose translations are exactly those in $\langle \tau_1, \tau_2 \rangle$ where if $\tau_1 = \tau_{A,B}$ and $\tau_2 = \tau_{A,C}$ then A, B, C are noncollinear points. The **translation lattice** for $\mathcal{W}$ determined by point P is the set of all images of P under the translations in $\mathcal{W}$. Since every translation in wallpaper group $\mathcal{W}$ is of the form $\tau_2^j \tau_1^i$, then all points A_{ij} form a translation lattice where $A_{ij} = \tau_2^j \tau_1^i(A)$. See Figure 11.1. A **unit cell** for $\mathcal{W}$ with respect to point A and generating translations τ_1 and τ_2 is a quadrilateral region with vertices A_{ij}, $A_{i+1, j}$, $A_{i, j+1}$, and $A_{i+1, j+1}$. A unit cell is always a quadrilateral region determined by a parallelogram. A translation lattice with a rectangular unit cell is called **rectangular**; a translation lattice with a rhombic unit cell is called **rhombic**. Our first task is to show that a translation lattice is necessarily rhombic or rectangular when $\mathcal{W}$ contains odd isometries.

Suppose σ_l is in wallpaper group $\mathcal{W}$. We wish to show that l is parallel to a diagonal of a rhombic unit cell or that l is parallel to a side of a rectangular unit cell. Let A be a point on l. Let $\tau_{A,P}$ be a shortest nonidentity translation in $\mathcal{W}$. There are two cases. Case 1: Neither $\overleftrightarrow{AP} = l$ nor $\overleftrightarrow{AP} \perp l$. Let $Q = \sigma_l(P)$. Then $\tau_{A,Q}$ is in $\mathcal{W}$ as $\tau_{A,Q} = \sigma_l \tau_{A,P} \sigma_l^{-1}$. See Figure 11.2. Since $AP = AQ$ and points A, P, Q are not collinear, then $\langle \tau_{A,P}, \tau_{A,Q} \rangle$ is the group of all

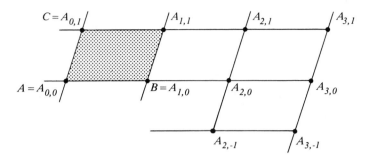

Figure 11.1

translations in $\mathscr{W}$ and l contains a diagonal of a rhombic unit cell. Case 2: $\overleftrightarrow{AP} = l$ or $\overleftrightarrow{AP} \perp l$. Let a be perpendicular to $\overleftrightarrow{AP}$ at A; let m be the perpendicular bisector of $\overline{AP}$; and let $n = \sigma_a(m)$. See Figure 11.3. Let $\tau_{A,R}$ be a shortest translation in $\mathscr{W}$ that is not in $\langle \tau_{A,P} \rangle$. Then R is on m, on n, or between m and n, as otherwise $\tau_{A,P}^{\pm 1}\tau_{A,R}$ is shorter than $\tau_{A,R}$. Further, considering both $\tau_{A,R}$ and its inverse, we may suppose without loss of generality that $\tau_{A,R}$ is such that R is on m, on a, or between m and a. Let $S = \sigma_l(R)$. Assume R is between m and a. If $l = \overleftrightarrow{AP}$, then $\tau_{A,S}\tau_{A,R}$ is a translation in $\mathscr{W}$ shorter than $\tau_{A,P}$; if $l \perp \overleftrightarrow{AP}$, then $\tau_{S,A}\tau_{A,R}$ is a translation in $\mathscr{W}$ shorter than $\tau_{A,P}$. Therefore, we must have R is on m or on a. If R is on m, then $\langle \tau_{A,R}, \tau_{A,S} \rangle$ is the same as $\langle \tau_{A,P}, \tau_{A,R} \rangle$ since $\tau_{A,S}\tau_{A,R} = \tau_{A,P}$. Thus l is parallel to a diagonal of a rhombic unit cell ($\square ARPS_1$ in Figure 11.3) with respect to point A and generating translations $\tau_{A,R}$ and $\tau_{A,S}$. On the other hand, if R is on a, then $\langle \tau_{A,P}, \tau_{A,R} \rangle$ is the group of all translations in $\mathscr{W}$ and l is parallel to a side of a rectangular unit cell for $\mathscr{W}$. This finishes the proof of the following theorem.

Theorem 11.1. *If σ_l is in wallpaper group $\mathscr{W}$, then l is parallel to a diagonal of a rhombic unit cell for $\mathscr{W}$ or else l is parallel to a side of a rectangular unit cell for $\mathscr{W}$.*

Now suppose wallpaper group $\mathscr{W}$ contains no reflections but does contain a glide reflection with axis line l. By the comments following Theorem 10.2,

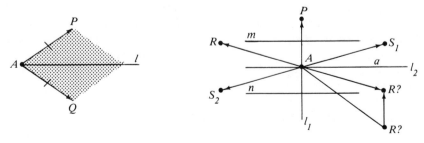

Figure 11.2   Figure 11.3

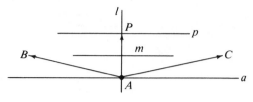

Figure 11.4

we see that the smallest group containing the glide reflection and the trans-
lations in $\mathcal{W}$ that fix l is a frieze group $\mathcal{F}_l^3$ generated by glide reflection γ
with axis l. Hence, we may suppose γ^2 is a shortest translation fixing l. Let
A be a point on l. Let a be perpendicular to l at A, $m = \gamma(a)$, $p = \gamma^2(a)$, and
$P = \gamma^2(A)$. So $\tau_{A, P}$ is a shortest translation in $\langle \gamma \rangle$. Let $\tau_{A, B}$ be a shortest
translation in $\mathcal{W}$ that is not in $\langle \gamma^2 \rangle$. Since $\tau_{A, P}^{\pm 1} \tau_{A, B}$ cannot be shorter than
$\tau_{A, B}$, we may suppose without loss of generality that B is on a or lies between
a and p. If B is on a, then $\mathcal{W}$ has a rectangular translation lattice and l is
parallel to a side of a rectangular unit cell. Suppose B is between a and p.
See Figure 11.4. Let $C = \sigma_l(B)$. Then $\tau_{A, C}$ is in $\mathcal{W}$ as $\tau_{A, C} = \gamma \tau_{A, B} \gamma^{-1}$. So
$\tau_{A, C} \tau_{A, B} = \gamma^2$ and B is on m. So $\square ABPC$ is a rhombic unit cell with l containing a diagonal. Thus we have the following.

Theorem 11.2. *If wallpaper group $\mathcal{W}$ contains a glide reflection, then $\mathcal{W}$ has a
translation lattice that is rhombic or rectangular.*

If glide reflection γ takes point A to point P in the translation lattice
determined by A for wallpaper group $\mathcal{W}$, then γ followed by $\tau_{P, A}$ must be a
reflection as the product is an odd isometry fixing point A. In particular, this
proves the following.

Theorem 11.3. *If a glide reflection in wallpaper group $\mathcal{W}$ fixes a translation
lattice for $\mathcal{W}$, then $\mathcal{W}$ contains a reflection.*

This finishes the preliminary results needed about odd isometries in a
wallpaper group. We turn to the rotations. Point P is an *n-center* for a group
$\mathcal{G}$ of isometries if the rotations in $\mathcal{G}$ with center P form a finite cyclic group
C_n with $n > 1$. A *figure* is a nonempty set of points. If point P is an n-center
for the symmetry group for a figure, then P is also called an *n-center* for
the figure. A *center of symmetry* is an n-center for some n. So, if point P is a
4-center for some figure, then P is a point of symmetry for that figure since
$\sigma_P = \rho_{P, 90}^2$. In this case, point P is a point of symmetry but P is not a 2-
center. Also note that if point Q is a 3-center for some figure then Q is not a
point of symmetry for that figure.

Examining the n-centers for a group turns out to be very profitable. First,
for a given n, the set of n-centers must be fixed by every isometry in the group.
To see this, suppose $\alpha(P) = Q$ for some isometry α in group $\mathcal{G}$. From the

equations $\alpha \rho_{P,\Theta} \alpha^{-1} = \rho_{Q,\pm\Theta}$ and $\alpha^{-1} \rho_{Q,\Phi} \alpha = \rho_{P,\pm\Phi}$, we see that Q is an n-center iff P is an n-center (for the same n). This very important analogue of Theorem 10.1 is stated below, where the second part of that theorem is repeated here for emphasis.

Theorem 11.4. *For given n, if point P is an n-center for group $\mathcal{G}$ of isometries and $\mathcal{G}$ contains an isometry that takes P to Q, then Q is an n-center for $\mathcal{G}$. If l is a line of symmetry for a figure and the symmetry group for the figure contains an isometry that takes l to m, then m is a line of symmetry for the figure.*

The first three exercises in Section 11.3 use Theorem 11.4 and probably should be done before starting the next section. In fact, the first nine of the exercises use this important theorem, and all nine could be done now if the groups denoted are assumed to be defined by the first equation in corresponding Figures 11.8, 11.9, 11.12, 11.14, 11.15, 11.18–11.20, and 11.22–11.30 from the next section.

Since a closest center of symmetry to a given n-center is often needed in the proofs below, we want to show that certain centers of symmetry cannot be arbitrarily close. Suppose rotations $\rho_{A, 360/n}$ and $\rho_{P, 360/n}$ with $P \neq A$ and $n > 1$ are in wallpaper group $\mathcal{W}$. Then $\mathcal{W}$ contains the product $\rho_{P, 360/n} \rho_{A, -360/n}$, which is a nonidentity translation $\tau_2^j \tau_1^i$ for some i and j by the Angle-addition Theorem. So

$$\rho_{P, 360/n} = \tau_2^j \tau_1^i \rho_{A, 360/n} \quad \text{and} \quad \rho_{P, 360/n}(A) = \tau_2^j \tau_1^i \rho_{A, 360/n}(A) = A_{ij}.$$

Hence, either P is the midpoint of A and A_{ij} (when $n = 2$) or else $\triangle APA_{ij}$ is isosceles. In either case, $2AP = AP + PA_{ij} \geq AA_{ij} > 0$ by the triangle inequality. Therefore, $2AP$ is not less than the length of any nonidentity translation in $\mathcal{W}$.

Theorem 11.5. *If $\rho_{A, 360/n}$ and $\rho_{P, 360/n}$ with $P \neq A$ and $n > 1$ are in wallpaper group $\mathcal{W}$, then $2AP$ is not less than the length of the shortest nonidentity translation in $\mathcal{W}$.*

The theorem above more precisely states—but contains—each of the following:

(1) No two n-centers (same n) can be "too close."
(2) A 2-center and a 4-center can't be too close.
(3) A 3-center and a 6-center can't be too close.
(4) A 2-center and a 6-center can't be too close.

We shall use (1) immediately to show that the possible values of n such that there is an n-center in a wallpaper group are rather restricted.

Suppose point P is an n-center of wallpaper group $\mathcal{W}$. Let Q be an n-center (same n) at the least possible distance from P with $Q \neq P$. The existence of point Q is assured by the previous two theorems. Let $R = \rho_{Q, 360/n}(P)$.

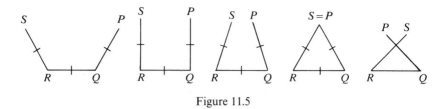

Figure 11.5

Then R is an n-center and $PQ = QR$. Let $S = \rho_{R, 360/n}(Q)$. Then S is an n-center and $RQ = RS$. If $S = P$, then $n = 6$. See Figure 11.5. If $S \neq P$, then we must have $SP \geq PQ = RQ$ by the choice of Q. Hence, if $S \neq P$, then $n \leq 4$. So n is one of 2, 3, 4, or 6. We have proved the **Crystallographic Restriction**:

Theorem 11.6. *If point P is an n-center for a wallpaper group, then n is one of 2, 3, 4, or 6.*

As an immediate corollary of the Crystallographic Restriction we have the following.

Theorem 11.7. *If a wallpaper group contains a 4-center, then the group contains neither a 3-center nor a 6-center.*

This follows from the fact that both $\rho_{P, 120}$ and $\rho_{Q, 90}$ cannot be in the same wallpaper group because the product $\rho_{P, 120}\rho_{Q, -90}$ is a rotation of $30°$ about some point and cannot be in any wallpaper group by the Crystallographic Restriction.

§11.2 Wallpaper Groups and Patterns

We shall find all possible wallpaper groups $\mathscr{W}$, beginning with those groups that contain an n-center. By the Crystallographic Restriction, it is sufficient to consider only values 6, 3, 4, and 2 for n. We begin by proving the following theorem which shows the abundance of symmetry required to support a 6-center.

Theorem 11.8. *Suppose A is a 6-center for wallpaper group $\mathscr{W}$. There there are no 4-centers for $\mathscr{W}$. Further, the center of symmetry nearest to A is a 2-center M, and A is the center of a regular hexagon whose vertices are 3-centers and whose sides are bisected by 2-centers. All the centers of symmetry for $\mathscr{W}$ are determined by A and M.*

To begin the proof, we note that $\mathscr{W}$ can contain no 4-centers since $\mathscr{W}$ contains the 6-center A (Theorem 11.7). Let M be an n-center nearest to A. If M were a 3-center or a 6-center, then there would be a center F closer to A

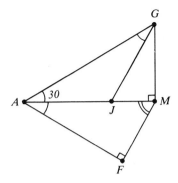

Figure 11.6

than M, where $\rho_{M,120}\rho_{A,60} = \rho_{F,180}$. See Figure 11.6 above. Hence, M must be a 2-center. Define point G by the equation $\rho_{M,180}\rho_{A,-60} = \rho_{G,120}$. So G is either a 3-center or a 6-center. However, G cannot be a 6-center as then there would be a center J between A and M, where J is defined by the equation $\rho_{G,60}\rho_{A,60} = \rho_{J,120}$. Hence G must be a 3-center. The images of G under the powers of $\rho_{A,60}$ are the vertices of the hexagon in the statement of the theorem. With $B = \sigma_M(A)$ and $C = \rho_{A,60}(B)$, then B and C are 6-centers for $\mathcal{W}$. The centers of symmetry determined by the 6-center A and the 2-center M are as arranged in Figure 11.7. With $N = \rho_{A,60}(M)$, then N is a 2-center for $\mathcal{W}$. Also, since 6-center A must go to a 6-center under an element of $\mathcal{W}$, then $\sigma_M\sigma_A$ and $\sigma_N\sigma_A$ are shortest translations in $\mathcal{W}$. Hence $\tau_{A,B}$ and $\tau_{A,C}$ must generate the translation subgroup of $\mathcal{W}$. We have our first specific wallpaper

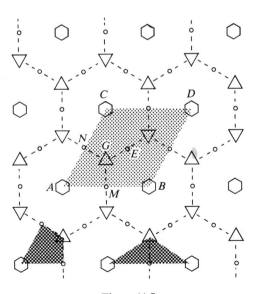

Figure 11.7

group: $\mathscr{W}_6 = \langle \tau_{A,B}, \tau_{A,C}, \rho_{A,60} \rangle = \langle \rho_{A,60}, \sigma_M \rangle$ where $\triangle ABC$ is equilateral and M is the midpoint of $\overline{AB}$.

In Figure 11.7, a unit cell determined by $\square ABDC$ is lightly shaded with M the midpoint of $\overline{AB}$ and N the midpoint of $\overline{AC}$. This notation will be used throughout, with E the point such that $\square NAME$ is a parallelogram. The two darker regions in Figure 11.7 are called *bases* for $\mathscr{W}_6$. A smallest polygonal region t such that the plane is covered by $\{\alpha(t) | \alpha \in \mathscr{W}\}$ is called a (polygonal) **base** for wallpaper group $\mathscr{W}$. (Fancier bases are possible but not necessary for our purpose here.) The bases may be used to create wallpaper patterns having a given wallpaper group as symmetry group. If t' is a figure with identity symmetry group in base t, then the union of all images $\alpha(t')$ with α in $\mathscr{W}$ is a figure with all the symmetries in $\mathscr{W}$. This figure is said to have **motif** t'. In producing a wallpaper pattern, once motif t' is picked for base t, that part of the pattern in a unit cell is determined and this is then just translated throughout the plane to give the wallpaper pattern. For each wallpaper group, a pattern will be given below where the motif is the same check used to illustrate the frieze patterns in Figures 10.1 through 10.7. For each of the seventeen wallpaper groups we shall have a figure such as Figure 11.8. These will have a pattern with the check motif, a unit cell with a base and symmetries indicated, two sets of generators for the group, and two designations for the

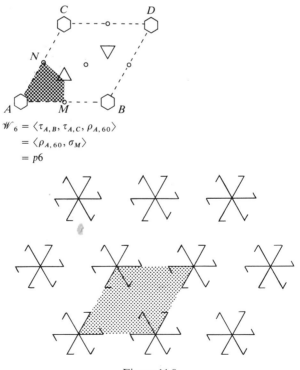

Figure 11.8

group. In addition to the notation of Fejes Tóth involving "$\mathcal{W}$," the *short international form* used by crystallographers will be given. For example, group $\mathcal{W}_6$ is designated $p6$ by crystallographers.

Consider extending $\mathcal{W}_6$ to $\mathcal{W}$. Since the rhombic translation lattice of 6-centers determined by 6-center A must be fixed by any isometry in $\mathcal{W}$, then (Theorem 11.3) any extensions of $\mathcal{W}_6$ are obtained by adding reflections that fix this translation lattice. However, because of the richness of $\mathcal{W}_6$, adding any one of the possible reflections requires introducing all of the possible reflections. See Figure 11.9. Let $\mathcal{W}_6^1 = \langle \tau_{A,B}, \tau_{A,C}, \rho_{A,60}, \sigma_{\overleftrightarrow{MC}} \rangle$. From the result on $\mathcal{W}_6$, then $\mathcal{W}_6^1 = \langle \rho_{A,60}, \sigma_M, \sigma_{\overleftrightarrow{MC}} \rangle$. So

$$\mathcal{W}_6^1 = \langle \sigma_{\overleftrightarrow{AG}}, \sigma_{\overleftrightarrow{GM}}, \sigma_{\overleftrightarrow{MA}} \rangle$$

and $\mathcal{W}_6^1$ is generated by the three reflections in the three lines that contain the sides of a 30°–60°–90° triangle. We leave it as an exercise to show that $\mathcal{W}_6^1 = \langle \rho_{A,60}, \sigma_{\overleftrightarrow{MC}} \rangle$. *A wallpaper pattern having symmetry group $\mathcal{W}_6$ has a 6-center but no line of symmetry; a wallpaper pattern having symmetry group $\mathcal{W}_6^1$ has a 6-center and a line of symmetry.* A wallpaper pattern having a 6-center has a symmetry group $\mathcal{W}_6$ or $\mathcal{W}_6^1$.

We turn to wallpaper groups with a 3-center but no 6-center and prove the following analogue to the previous theorem.

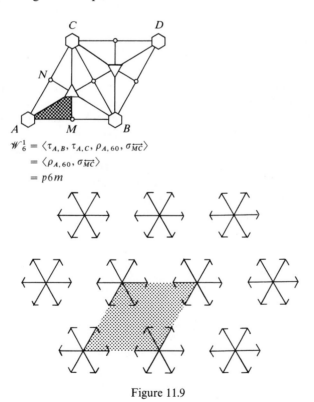

$$\mathcal{W}_6^1 = \langle \tau_{A,B}, \tau_{A,C}, \rho_{A,60}, \sigma_{\overleftrightarrow{MC}} \rangle$$
$$= \langle \rho_{A,60}, \sigma_{\overleftrightarrow{MC}} \rangle$$
$$= p6m$$

Figure 11.9

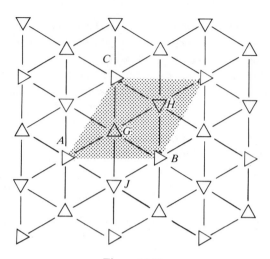

Figure 11.10

Theorem 11.9. *If A is a 3-center for wallpaper group $\mathcal{W}$ and there are no 6-centers for $\mathcal{W}$, then every center of symmetry for $\mathcal{W}$ is a 3-center and A is the center of a regular hexagon whose vertices are 3-centers. All the centers of symmetry for $\mathcal{W}$ are determined by A and a nearest 3-center.*

That every center P for $\mathcal{W}$ must be a 3-center follows from the fact that $\rho_{A,-120}\rho_{P,180}$ cannot be in $\mathcal{W}$ for any point P since $\mathcal{W}$ contains no 6-centers. Let G be a nearest 3-center to A. Let J be such that $\rho_{G,120}\rho_{A,120} = \rho_{J,240}$. Then J is a 3-center and $\triangle AGJ$ is an equilateral triangle. The images of G and J under powers of $\rho_{A,120}$ are the vertices of the hexagon in the statement of the theorem. Repetition of the argument for each 3-center shows that all the 3-centers are arrayed as in Figure 11.10 and finishes the proof of the theorem. Further, from Figure 11.11, we see that each of $\rho_{G,120}\tau_{A,G}$ and $\rho_{J,120}\tau_{A,J}$ is $\rho_{Q,120}$ where Q is the centroid of $\triangle AGJ$. Hence neither $\tau_{A,G}$ nor $\tau_{A,J}$ is in $\mathcal{W}$ as otherwise Q would be a 3-center nearer to A than G. So, if $\tau_{A,B}$ is a shortest translation in $\mathcal{W}$, then 3-center B is not a vertex of the hexagon of nearest 3-centers to A. Let B and C be defined by $\tau_{A,B} = \rho_{G,120}\rho_{A,-120}$ and

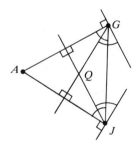

Figure 11.11

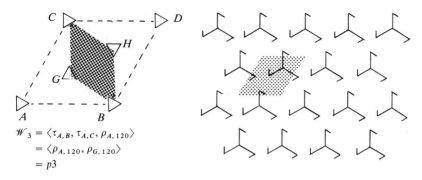

$$\mathcal{W}_3 = \langle \tau_{A,B}, \tau_{A,C}, \rho_{A,120} \rangle$$
$$= \langle \rho_{A,120}, \rho_{G,120} \rangle$$
$$= p3$$

Figure 11.12

$\tau_{A,C} = \rho_{G,-120}\rho_{A,120}$. Then $\tau_{A,B}$ and $\tau_{A,C}$ are shortest translations in $\mathcal{W}$ and take A to next nearest 3-centers. Hence, $\tau_{A,B}$ and $\tau_{A,C}$ generate the translation group of $\mathcal{W}$. Let $\mathcal{W}_3 = \langle \tau_{A,B}, \tau_{A,C}, \rho_{A,120} \rangle = \langle \rho_{A,120}, \rho_{G,120} \rangle$. If $\mathcal{W}$ contains no odd isometries then $\mathcal{W}$ must be $\mathcal{W}_3$. See Figure 11.12.

Consider extending $\mathcal{W}_3$ to a wallpaper group $\mathcal{W}$ without reflections by adding glide reflections. Then (Theorem 11.3), $\mathcal{W}$ must have a glide reflection that takes 3-center A to a 3-center that is not in the translation lattice determined by A. By composing this glide reflection with a translation and possibly a rotation about A, we may assume $\mathcal{W}$ contains a glide reflection that takes A to either G or J. Suppose γ is a glide reflection in $\mathcal{W}$ that takes A to G. Then $\gamma = \sigma_z \sigma_Z$ where Z is the midpoint of A and G and z is some line through G. Since σ_Z fixes the set of all 3-centers, then σ_z must also fix the set of all 3-centers. See Figure 11.13. By composing σ_z with a rotation about G, we may suppose without loss of generality that z is either the perpendicular bisector of $\overline{JB}$ or $z = \overleftrightarrow{GJ}$. The first is impossible as otherwise $\overleftrightarrow{AG}$ is the axis of γ and $\tau_{B,A}\gamma^2$ is a translation in $\mathcal{W}$ of length AG and shorter than $\tau_{A,B}$. So $z = \overleftrightarrow{GJ}$. However, then $\rho_{G,-120}\gamma = \sigma_{\overleftrightarrow{AG}}\sigma_z\sigma_z\sigma_Z = \sigma_{\overleftrightarrow{ZJ}}$ and $\mathcal{W}$ contains the reflection in the perpendicular bisector of $\overline{AG}$. Likewise, the presence of a glide reflection taking A to J implies the reflection in the perpendicular

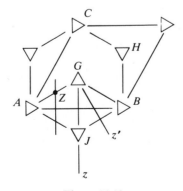

Figure 11.13

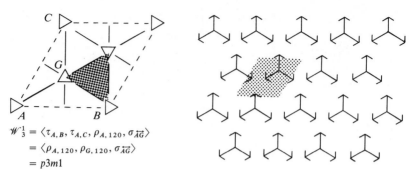

$$\mathscr{W}_3^1 = \langle \tau_{A,B}, \tau_{A,C}, \rho_{A,120}, \sigma_{\overleftrightarrow{AG}} \rangle$$
$$= \langle \rho_{A,120}, \rho_{G,120}, \sigma_{\overleftrightarrow{AG}} \rangle$$
$$= p3m1$$

Figure 11.14

bisector of $\overline{AJ}$ is in $\mathscr{W}$. In any case, $\mathscr{W}$ must contain a reflection if $\mathscr{W}$ is an extension of $\mathscr{W}_3$ and contains an odd isometry.

All extensions of $\mathscr{W}_3$ to a group with no 6-centers by adding odd isometries are obtained by adding reflections. If $\mathscr{W}_3$ is extended by adding σ_l, then line l must be a line of symmetry for the set of 3-centers. Since such a line must pass through at least one 3-center, we suppose l is a line through 3-center A. Let $\mathscr{W}_3^1 = \langle \tau_{A,B}, \tau_{A,C}, \rho_{A,120}, \sigma_{\overleftrightarrow{AG}} \rangle$ and $\mathscr{W}_3^2 = \langle \tau_{A,B}, \tau_{A,C}, \rho_{A,120}, \sigma_{\overleftrightarrow{AB}} \rangle$. See Figures 11.14 and 11.15. We leave as an exercise the proof that $\mathscr{W}_3^1$ is generated by the three reflections in the three lines containing the sides of an equilateral triangle. Groups $\mathscr{W}_3^1$ and $\mathscr{W}_3^2$ are obtained by adding to $\mathscr{W}_3$ the reflection in one of the diagonals of the rhombic unit cell determined by A. Adding the reflections in both diagonals would introduce a halfturn and a 6-center. So any wallpaper pattern containing only three centers has one of $\mathscr{W}_3$, $\mathscr{W}_3^1$, or $\mathscr{W}_3^2$ as its symmetry group. *A wallpaper pattern having symmetry groups $\mathscr{W}_3$ has a 3-center, has no 6-center, and has no line of symmetry. A wallpaper pattern having symmetry group $\mathscr{W}_3^1$ has a 3-center, has no 6-center, and every 3-center is on a line of symmetry. A wallpaper pattern having symmetry group $\mathscr{W}_3^2$ has a 3-center off a line of symmetry but no 6-center.*

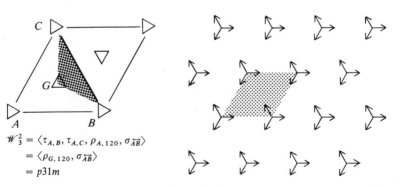

$$\mathscr{W}_3^2 = \langle \tau_{A,B}, \tau_{A,C}, \rho_{A,120}, \sigma_{\overleftrightarrow{AB}} \rangle$$
$$= \langle \rho_{G,120}, \sigma_{\overleftrightarrow{AB}} \rangle$$
$$= p31m$$

Figure 11.15

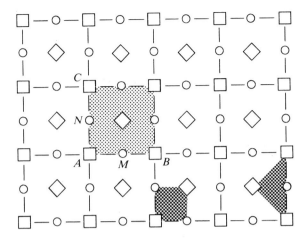

Figure 11.16

The short international forms p3m1 and p31m have often been interchanged in the mathematical literature and caution is advised whenever these are encountered.

The following theorem regarding 4-centers in a wallpaper group is analogous to that regarding 6-centers. The proof follows the statement of the theorem. See Figure 11.16.

Theorem 11.10. *Suppose A is a 4-center for wallpaper group $\mathcal{W}$. Then, there are no 3-centers for $\mathcal{W}$ and there are no 6-centers for $\mathcal{W}$. Further, the center of symmetry nearest to A is a 2-center M, and A is the center of a square whose vertices are 4-centers and whose sides are bisected by 2-centers. All the centers of symmetry for $\mathcal{W}$ are determined by A and M.*

By the corollary (Theorem 11.7) to the Crystallographic Restriction, if A is a 4-center for wallpaper group $\mathcal{W}$, then every center of symmetry for $\mathcal{W}$ is either a 2-center or a 4-center. Let M be a center of symmetry nearest to A. If M were a 4-center, then K would be a center of symmetry closer to A than M where K is given by $\rho_{M,90}\rho_{A,90} = \sigma_K$. See Figure 11.17. So M must be a 2-center. Then E is a 4-center where $\rho_{M,180}\rho_{A,-90} = \rho_{E,90}$. The images of E and M under the powers of $\rho_{A,90}$ are, respectively, the vertices and midpoints of the square in the statement of the theorem. Translation $\tau_{A,E}$ is not in $\mathcal{W}$ as otherwise Z is a center of symmetry closer to A than M where $\tau_{A,E}\sigma_A = \sigma_Z$. With $N = \rho_{A,90}(M)$, $\tau_{A,B} = \sigma_M\sigma_A$, and $\tau_{A,C} = \sigma_N\sigma_A$, then $\square NAME$ is a square and $\tau_{A,B}$ and $\tau_{A,C}$ are shortest translations in $\mathcal{W}$ and generate the translation subgroup. Thus, there is no more room for any more centers of symmetry than already accounted (Theorem 11.5). The centers of symmetry for $\mathcal{W}$ are as arranged in Figure 11.16. Let $\mathcal{W}_4 = \langle \tau_{A,B}, \tau_{A,C},$

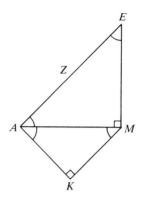

Figure 11.17

$\rho_{A, 90}\rangle$ where E is the center of square $\square ABDC$. If $\mathscr{W}$ contains no odd isometries then $\mathscr{W}$ must be $\mathscr{W}_4$. It should be easy to check that $\mathscr{W}_4$ is generated
by $\rho_{A, 90}$ and $\rho_{E, 90}$. See Figure 11.18.

Consider extending $\mathscr{W}_4$ to wallpaper group $\mathscr{W}$ by adding odd isometries.
If σ_l is in $\mathscr{W}$, then l must be a line of symmetry for the set of all 4-centers in $\mathscr{W}$.
Because of the abundance of rotations in $\mathscr{W}_4$, it is seen to be sufficient to
consider adding either a reflection in a line of symmetry through a 4-center
or else a reflection in a line of symmetry off all the 4-centers. Lines $\overleftrightarrow{AE}$ and
$\overleftrightarrow{MN}$ will serve our purpose. First, let $\mathscr{W}_4^1 = \langle \tau_{A, B}, \tau_{A, C}, \rho_{A, 90}, \sigma_{\overleftrightarrow{AE}} \rangle$. See
Figure 11.19. It is easy to check that $\mathscr{W}_4^1$ is also generated by the three re-
flections in the three lines that contain the sides of an isosceles right triangle.
Secondly, let $\mathscr{W}_4^2 = \langle \tau_{A, B}, \tau_{A, C}, \rho_{A, 90}, \sigma_{\overleftrightarrow{MN}} \rangle$. See Figure 11.20. Both $\sigma_{\overleftrightarrow{AE}}$
and $\sigma_{\overleftrightarrow{MN}}$ cannot be added to $\mathscr{W}_4$ without introducing a center of symmetry
closer to A than M.

To consider the possibility of extending $\mathscr{W}_4$ to a wallpaper group $\mathscr{W}$
without reflections by adding odd isometries, it is sufficient (Theorem 11.3)
to suppose $\mathscr{W}$ contains a glide reflection taking 4-center A to a 4-center that

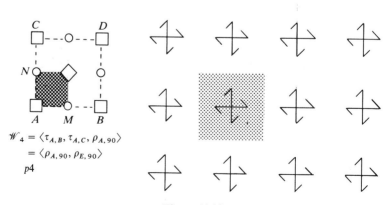

$\mathscr{W}_4 = \langle \tau_{A, B}, \tau_{A, C}, \rho_{A, 90} \rangle$
$\quad = \langle \rho_{A, 90}, \rho_{E, 90} \rangle$

$p4$

Figure 11.18

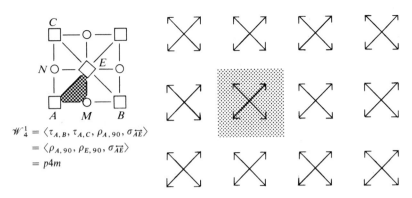

$$\mathcal{W}_4^1 = \langle \tau_{A,B}, \tau_{A,C}, \rho_{A,90}, \sigma_{\overrightarrow{AE}} \rangle$$
$$= \langle \rho_{A,90}, \rho_{E,90}, \sigma_{\overrightarrow{AE}} \rangle$$
$$= p4m$$

Figure 11.19

is not in the translation lattice determined by A. By composing this glide reflection with an appropriate translation, we may suppose $\mathcal{W}$ contains a glide reflection γ taking A to E. With Z the midpoint of A and E, then $\gamma = \sigma_z \sigma_Z$ for some line z through E. Since γ must fix the set of all 4-centers, then z must be one of $\overleftrightarrow{ME}$, $\overleftrightarrow{BE}$, or $\overleftrightarrow{NE}$. However, γ followed, respectively, by $\rho_{E,90}$, $\rho_{E,180}$, or $\rho_{E,270}$ gives the reflection in $\overleftrightarrow{MN}$. Extending $\mathcal{W}_4$ by odd isometries leads only to $\mathcal{W}_4^1$ or $\mathcal{W}_4^2$. *A wallpaper pattern having symmetry group $\mathcal{W}_4$ has a 4-center and no line of symmetry. A wallpaper pattern having symmetry group $\mathcal{W}_4^1$ has a line of symmetry on a 4-center. A wallpaper pattern having symmetry group $\mathcal{W}_4^2$ has a 4-center and a line of symmetry off all 4-centers.*

Now suppose wallpaper group $\mathcal{W}$ has a 2-center A and every center of symmetry for $\mathcal{W}$ is a 2-center. So σ_A is in $\mathcal{W}$. With $\langle \tau_{A,B}, \tau_{A,C} \rangle$ the translation subgroup of $\mathcal{W}$, let $\sigma_M = \tau_{A,B}\sigma_A$, $\sigma_N = \tau_{A,C}\sigma_A$, and $\sigma_E = \sigma_N \sigma_A \sigma_M$. Points M, N, E are 2-centers and we have our usual notation with $\square ABDC$ defining a unit cell. See Figure 11.21. Every point A_{ij} in the translation lattice determined by A is a 2-center as well as the midpoint of any two such lattice points. There can be no more centers of symmetry than these. Let $\mathcal{W}_2 = \langle \tau_{A,B}, \tau_{A,C}, \sigma_A \rangle$. If $\mathcal{W}$ contains no odd isometries, then $\mathcal{W} = \mathcal{W}_2$. See

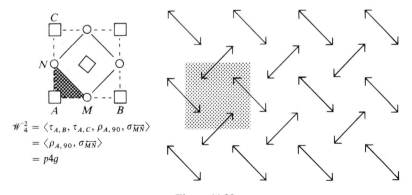

$$\mathcal{W}_4^2 = \langle \tau_{A,B}, \tau_{A,C}, \rho_{A,90}, \sigma_{\overrightarrow{MN}} \rangle$$
$$= \langle \rho_{A,90}, \sigma_{\overrightarrow{MN}} \rangle$$
$$= p4g$$

Figure 11.20

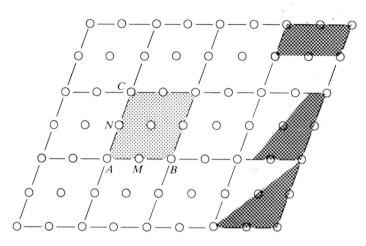

Figure 11.21

Figure 11.22. It is easy to check that $\mathscr{W}_2$ is generated by $\tau_{A,B}$, $\tau_{A,C}$, and σ_E and also by σ_M, σ_E, and σ_N.

Consider extending $\mathscr{W}_2$ to wallpaper group $\mathscr{W}$ by adding only odd isometries. Suppose σ_l is in $\mathscr{W}$. Then $\mathscr{W}$ has a rhombic or rectangular translation lattice (Theorem 11.1). In the nonrectangular rhombic case, line l is parallel to a diagonal of a unit cell and so must pass through a 2-center. In this case, we may suppose A to be a 2-center on l. Then l contains a diagonal unit cell of the translation lattice determined by A. However, adding the reflection in one diagonal of a unit cell necessitates adding the reflection in the other diagonal as the center of the unit cell is a 2-center. Let

$$\mathscr{W}_2^1 = \langle \tau_{A,B}, \tau_{A,C}, \sigma_{\overrightarrow{AE}}, \sigma_{\overrightarrow{BE}} \rangle.$$

See Figure 11.23. Therefore, in the nonrectangular case, we have only the one possibility that $\mathscr{W} = \mathscr{W}_2^1$. It should be easy to check that $\mathscr{W}_2^1$ is also generated by $\sigma_{\overrightarrow{AE}}$, $\sigma_{\overrightarrow{BE}}$, and σ_M. If a rhombic unit cell is rectangular, then the unit cell is square and is a special case of the general rectangular case considered next.

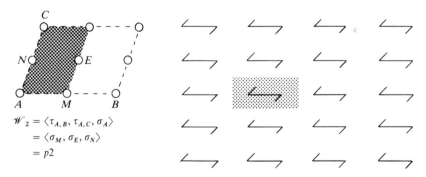

$\mathscr{W}_2 = \langle \tau_{A,B}, \tau_{A,C}, \sigma_A \rangle$
$\quad = \langle \sigma_M, \sigma_E, \sigma_N \rangle$
$\quad = p2$

Figure 11.22

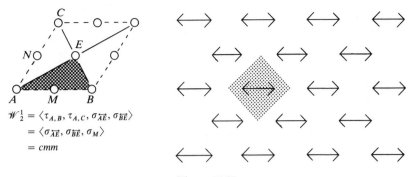

$$\mathcal{W}_2^1 = \langle \tau_{A,B}, \tau_{A,C}, \sigma_{\overline{AE}}, \sigma_{\overline{BE}} \rangle$$
$$= \langle \sigma_{\overline{AE}}, \sigma_{\overline{BE}}, \sigma_M \rangle$$
$$= cmm$$

Figure 11.23

An extension $\mathcal{W}$ of $\mathcal{W}_2$ cannot have a reflection in a diagonal of a unit cell unless the unit cell is rhombic and cannot have reflections in both a diagonal and a line parallel to a side since every n-center is a 2-center. Thus, to extend $\mathcal{W}_2$ with reflections there remains to consider only the case where a unit cell is rectangular (possibly square) and σ_l is in $\mathcal{W}$ with l parallel to a side of unit cell defined by $\square ABDC$. There are the two possibilities that either l passes through a 2-center or else l passes between two adjacent rows of 2-centers. In the first case, introducing the reflection in one of the lines that contains a side of $\square NAME$ requires the introduction of the reflection in each of these lines. In this case, $\mathcal{W}$ is $\mathcal{W}_2^2$ defined by

$$\mathcal{W}_2^2 = \langle \tau_{A,B}, \tau_{A,C}, \sigma_{\overleftrightarrow{AM}}, \sigma_{\overleftrightarrow{AN}} \rangle.$$

See Figure 11.24. In the second case, where l passes between two adjacent rows of 2-centers, we may suppose without loss of generality that l is parallel to $\overleftrightarrow{AN}$. In this case, $\mathcal{W}$ is $\mathcal{W}_2^3$ defined by $\mathcal{W}_2^3 = \langle \tau_{A,B}, \tau_{A,C}, \sigma_A, \sigma_p \rangle$ where p is the perpendicular bisector of $\overline{AM}$. See Figure 11.25. We have finished extending $\mathcal{W}_2$ by adding only reflections.

Consider extending $\mathcal{W}_2$ to wallpaper group $\mathcal{W}$ by adding a glide reflection γ such that no reflections are introduced. A glide reflection having an axis that passes through a 2-center necessitates the introduction of a reflection. Hence, the axis of γ must pass between two adjacent rows of 2-centers. This

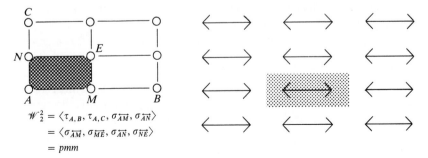

$$\mathcal{W}_2^2 = \langle \tau_{A,B}, \tau_{A,C}, \sigma_{\overleftrightarrow{AM}}, \sigma_{\overleftrightarrow{AN}} \rangle$$
$$= \langle \sigma_{\overleftrightarrow{AM}}, \sigma_{\overleftrightarrow{ME}}, \sigma_{\overleftrightarrow{AN}}, \sigma_{\overleftrightarrow{NE}} \rangle$$
$$= pmm$$

Figure 11.24

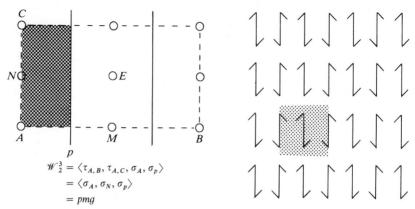

$$\mathcal{W}_2^3 = \langle \tau_{A,B}, \tau_{A,C}, \sigma_A, \sigma_p \rangle$$
$$= \langle \sigma_A, \sigma_N, \sigma_p \rangle$$
$$= pmg$$

Figure 11.25

requires that the parallelogram unit cell be rectangular. We shall see that one choice of axis for γ introduces glide reflections whose axes consist of all possible candidates. Let p be the perpendicular bisector of $\overline{AM}$; let q be the perpendicular bisector of $\overline{AN}$. A glide reflection with axis p taking 2-center M to 2-center C followed by $\tau_{C,A}$ produces the undesired reflection σ_p. So let γ be the glide reflection taking M to N and A to E. Let ε be the glide reflection taking N to M and A to E. The axis of γ is p and $\gamma^2 = \tau_{A,C}$; the axis of ε is q and $\varepsilon^2 = \tau_{A,B}$. Check that $\gamma\sigma_A = \varepsilon$. Let $\mathcal{W}_2^4 = \langle \tau_{A,B}, \tau_{A,C}, \sigma_A, \gamma \rangle$. Then $\mathcal{W}_2^4$ not only contains both γ and ε—and hence all possible glide reflections whose presence does not also require reflections—but $\mathcal{W}_2^4$ is even generated by γ and ε. See Figure 11.26, where the axes of glide reflections are indicated by broken lines. Thus the extension of $\mathcal{W}_2$ to a wallpaper group by adding only odd isometries gives one of $\mathcal{W}_2^1$, $\mathcal{W}_2^2$, $\mathcal{W}_2^3$, or $\mathcal{W}_2^4$.

A wallpaper pattern having symmetry group $\mathcal{W}_2$ has a 2-center, every center of symmetry is a 2-center, and is not fixed by any odd isometry. A wallpaper pattern having symmetry group $\mathcal{W}_2^1$ has a 2-center, every center of symmetry is a 2-center, and some but not all 2-centers are on a line of symmetry. A wallpaper pattern having symmetry group $\mathcal{W}_2^2$ has a 2-center, every center of symmetry is a 2-center, and every 2-center is on a line of symmetry. A wallpaper pattern having symmetry group $\mathcal{W}_2^3$ has a 2-center, every center of symmetry is a 2-center, has a line of symmetry, and all lines of symmetry are parallel. A wallpaper pattern having symmetry group $\mathcal{W}_2^4$ has a 2-center, every center of symmetry is a 2-center, has no line of symmetry, but is fixed by a glide reflection.

At last we come to those wallpaper groups $\mathcal{W}$ that have no center of symmetry. If $\mathcal{W}$ contains σ_l, suppose A is on l. If $\mathcal{W}$ contains no reflection but does have a glide reflection, suppose A is on the axis of a glide reflection in $\mathcal{W}$. There is the case where $\mathcal{W}$ contains no isometries other than translations. In this case, the $\mathcal{W} = \mathcal{W}_1$ where $\mathcal{W}_1 = \langle \tau_{A,B}, \tau_{A,C} \rangle$ with A arbitrary and A, B, C noncollinear. See Figure 11.27.

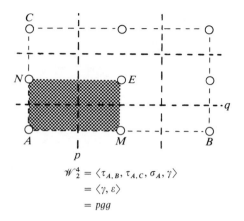

$$\mathscr{W}_2^4 = \langle \tau_{A,B}, \tau_{A,C}, \sigma_A, \gamma \rangle$$
$$= \langle \gamma, \varepsilon \rangle$$
$$= pgg$$

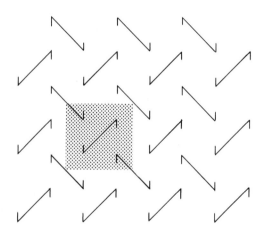

Figure 11.26

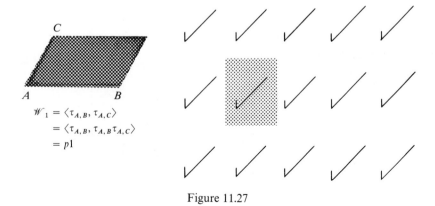

$$\mathscr{W}_1 = \langle \tau_{A,B}, \tau_{A,C} \rangle$$
$$= \langle \tau_{A,B}, \tau_{A,B}\tau_{A,C} \rangle$$
$$= p1$$

Figure 11.27

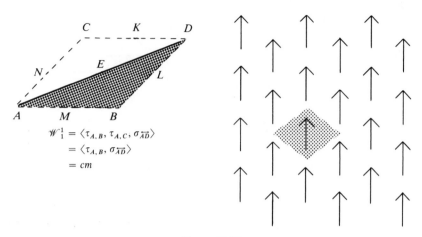

$$\mathscr{W}_1^1 = \langle \tau_{A,B}, \tau_{A,C}, \sigma_{\overleftrightarrow{AD}} \rangle$$
$$= \langle \tau_{A,B}, \sigma_{\overleftrightarrow{AD}} \rangle$$
$$= cm$$

Figure 11.28

Consider extending $\mathscr{W}_1$ to wallpaper group $\mathscr{W}$ by adding only odd isometries. If σ_l is in $\mathscr{W}$ with A on l, then (Theorem 11.1) there is a rhombic unit cell defined by $\square ABDC$ with $l = \overleftrightarrow{AD}$ or else there is a rectangular unit cell defined by $\square ABDC$ with $l = \overleftrightarrow{AC}$. In case $\square ABDC$ is a square, $\mathscr{W}$ cannot have reflections in both $\overleftrightarrow{AD}$ and $\overleftrightarrow{AC}$ because $\mathscr{W}$ contains no rotations. Let $\mathscr{W}_1^1 = \langle \tau_{A,B}, \tau_{A,C}, \sigma_{\overleftrightarrow{AD}} \rangle$ and $\mathscr{W}_1^2 = \langle \tau_{A,B}, \tau_{A,C}, \sigma_{\overleftrightarrow{AC}} \rangle$. If $\mathscr{W}$ contains a reflection then $\mathscr{W}$ is one of $\mathscr{W}_1^1$ or $\mathscr{W}_1^2$. See Figure 11.28 and note that $\overleftrightarrow{NK}$ is not a line of symmetry for $\mathscr{W}_1^1$. Let γ be the glide reflections with axis $\overleftrightarrow{NK}$ that takes N to K. So $\gamma^2 = \tau_{A,D}$. Check that γ is in $\mathscr{W}_1^1$ and, in fact, $\mathscr{W}_1^1$ is generated by γ and $\sigma_{\overleftrightarrow{AD}}$. On the other hand, all the glide reflections in $\mathscr{W}_1^2$ are of the form $(\sigma_{\overleftrightarrow{AC}} \tau_{A,B}^j) \tau_{A,C}^i$ with $i \neq 0$ and have axes that are also lines of symmetry for $\mathscr{W}_1^2$. This property can be used to distinguish patterns with symmetry groups $\mathscr{W}_1^1$ and $\mathscr{W}_1^2$. See Figure 11.29.

Finally, consider extending $\mathscr{W}_1$ with glide reflections only. The axes of the glide reflections must be parallel because $\mathscr{W}_1$ contains no rotations. We have

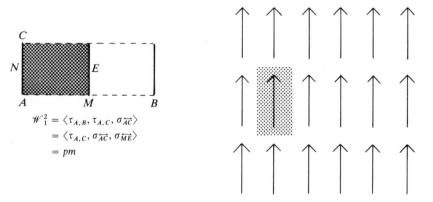

$$\mathscr{W}_1^2 = \langle \tau_{A,B}, \tau_{A,C}, \sigma_{\overleftrightarrow{AC}} \rangle$$
$$= \langle \tau_{A,C}, \sigma_{\overleftrightarrow{AC}}, \sigma_{\overleftrightarrow{ME}} \rangle$$
$$= pm$$

Figure 11.29

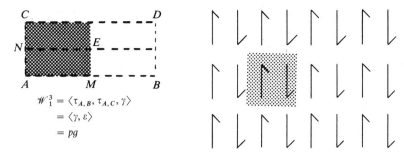

$$\mathscr{W}_1^3 = \langle \tau_{A,B}, \tau_{A,C}, \gamma \rangle$$
$$= \langle \gamma, \varepsilon \rangle$$
$$= pg$$

Figure 11.30

the one further case where one of the generating translations is the square of a glide reflection in $\mathscr{W}$. Let $\mathscr{W}_1^3 = \langle \tau_{A,B}, \tau_{A,C}, \gamma \rangle$ where γ is the glide reflection with axis $\overleftrightarrow{AB}$ that takes A to M. See Figure 11.30. So $\gamma^2 = \tau_{A,B}$. Clearly $\mathscr{W}_1^3$ is generated by $\tau_{A,C}$ and γ. Let $\varepsilon = \tau_{A,C}\gamma$. Check that ε is the glide reflection with axis $\overleftrightarrow{NE}$ that takes N to E and that $\mathscr{W}_1^3$ is generated by γ and ε. So $\mathscr{W}_1^3$ contains all possible glide reflections. The extension of $\mathscr{W}_1$ to a wallpaper group by adding only odd isometries gives one of $\mathscr{W}_1^1$, $\mathscr{W}_1^2$, or $\mathscr{W}_1^3$.

A wallpaper pattern having symmetry group $\mathscr{W}_1$ has no center of symmetry and is not fixed by any odd isometry. A wallpaper pattern having symmetry group $\mathscr{W}_1^1$ has no center of symmetry, is fixed by both reflections and glide reflections, but some axes of the glide reflections are not lines of symmetry. A wallpaper pattern having symmetry group $\mathscr{W}_1^2$ has no center of symmetry, is fixed by both reflections and glide reflections, and all axes of the glide reflections are lines of symmetry. A wallpaper pattern having symmetry group $\mathscr{W}_1^3$ has no center of symmetry, has no line of symmetry, but is fixed by a glide reflection.

Our search is at an end. All the wallpaper groups have been listed.

Theorem 11.11. *If $\mathscr{W}$ is a wallpaper group, then there are points and lines such that $\mathscr{W}$ is one of the seventeen groups*

$\mathscr{W}_1$	$\mathscr{W}_2$	$\mathscr{W}_4$	$\mathscr{W}_3$	$\mathscr{W}_6$
$\mathscr{W}_1^1$	$\mathscr{W}_2^1$	$\mathscr{W}_4^1$	$\mathscr{W}_3^1$	$\mathscr{W}_6^1$
$\mathscr{W}_1^2$	$\mathscr{W}_2^2$	$\mathscr{W}_4^2$	$\mathscr{W}_3^2$	
$\mathscr{W}_1^3$	$\mathscr{W}_2^3$			
	$\mathscr{W}_2^4$			

defined above.

Table 11.1 gives the key to the wallpaper patterns that we have been developing throughout the argument, where here the words *some* and *all* are to imply the existence of at least one, while *some* excludes at least one. Figure 11.31 gives a pattern for each group in corresponding position to the key. The patterns in this figure are called tilings and are the subject of the

Table 11.1 A key to the wallpaper patterns.

No n-centers	Only 2-centers	4-centers	Only 3-centers	6-centers
$\mathscr{W}_1$ $p1$ No odd isometries	$\mathscr{W}_2$ $p2$ No odd isometries	$\mathscr{W}_4$ $p4$ No lines of symmetry	$\mathscr{W}_3$ $p3$ No lines of symmetry	$\mathscr{W}_6$ $p6$ No lines of symmetry
$\mathscr{W}_1^1$ cm Some axes of glide reflection are not lines of symmetry	$\mathscr{W}_2^1$ cmm Some 2-centers not on a line of symmetry	$\mathscr{W}_4^1$ $p4m$ A line of symmetry on a 4-center	$\mathscr{W}_3^1$ $p3m1$ All 3-centers on a line of symmetry	$\mathscr{W}_6^1$ $p6m$ A line of symmetry
$\mathscr{W}_1^2$ pm All axes of glide reflection are lines of symmetry	$\mathscr{W}_2^2$ pmm All 2-centers on a line of symmetry	$\mathscr{W}_4^2$ $p4g$ A line of symmetry off the 4-centers	$\mathscr{W}_3^2$ $p31m$ A 3-center off all lines of symmetry	
$\mathscr{W}_1^3$ pg No line of symmetry but glide reflections	$\mathscr{W}_2^3$ pmg All lines of symmetry are parallel			
	$\mathscr{W}_2^4$ pgg No line of symmetry but glide reflections			

Figure 11.31

next chapter. Figure 11.32 gives a pattern for each of the seventeen wallpaper groups for practice using the key.

The classification theorem above for the wallpaper groups is known as **Fedorov's Theorem** as Fedorov first treated these groups in 1891, a few

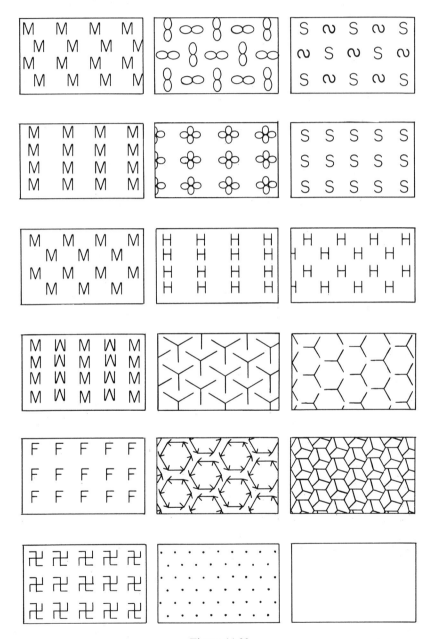

Figure 11.32

months after he gave the analogous three-dimensional groups. The theorem was rediscovered in 1897 by Fricke and Klein and again in 1924 by Polya and by Niggli. All seventeen groups were implicitly known to the Moors as illustrated in their ornament. In his famous book on group theory, Speiser considers this intuitive discovery of the ornamental groups one of the greatest mathematical achievements before modern times.

§11.3 Exercises

11.1. If T is the center of square $\square PQRS$, show that $\sigma_P, \sigma_Q, \sigma_R$ are in $\langle \sigma_{\overleftrightarrow{PQ}}, \sigma_S, \sigma_T \rangle$.

11.2. Show $\langle \rho_{A,60}, \sigma_{\overleftrightarrow{CG}} \rangle = \langle \sigma_{\overleftrightarrow{AG}}, \sigma_{\overleftrightarrow{CG}}, \sigma_{\overleftrightarrow{AB}} \rangle$ if G is the center of equilateral $\triangle ABC$.

11.3. Show $\langle \rho_{A,60}, \sigma_M \rangle = \langle \rho_{A,60}, \rho_{B,60} \rangle$ if M is the midpoint of $\overline{AB}$.

11.4. Show $\mathscr{W}_6^1 = \langle \rho_{A,60}, \sigma_{\overleftrightarrow{MC}} \rangle$ with notation as in Figure 11.9.

11.5. Show $\mathscr{W}_3^2 = \langle \rho_{G,120}, \sigma_{\overleftrightarrow{AB}} \rangle$ with notation as in Figure 11.15. Also, show $\mathscr{W}_3^2$ is generated by two glide reflections.

11.6. Show $\mathscr{W}_3^1$ is generated by the reflections in the lines containing the sides of an equilateral triangle.

11.7. With the notation as in Figures 11.18, 11.19, and 11.20, show that $\mathscr{W}_4$ is generated by $\rho_{A,90}$ and σ_M and also by $\rho_{A,90}$ and $\rho_{E,90}$, that $\mathscr{W}_4^1$ is generated by the three reflections in three lines containing the sides of an isosceles right triangle, and that $\mathscr{W}_4^2$ is generated by $\rho_{A,90}$ and $\sigma_{\overleftrightarrow{MN}}$.

11.8. Using the notation in the text, show that $\mathscr{W}_2$ is generated by $\tau_{A,B}, \tau_{A,C}$ and σ_E and by σ_M, σ_E, and σ_N, that $\mathscr{W}_2^1$ is generated by $\sigma_{\overleftrightarrow{AE}}, \sigma_{\overleftrightarrow{BE}}$, and σ_M, that $\mathscr{W}_2^2$ is generated by $\sigma_{\overleftrightarrow{AM}}, \sigma_{\overleftrightarrow{ME}}, \sigma_{\overleftrightarrow{AN}}$, and $\sigma_{\overleftrightarrow{NE}}$, that $\mathscr{W}_2^3$ is generated by σ_A, σ_N, and σ_p, and that $\mathscr{W}_2^4$ is generated by two glide reflections with perpendicular axes.

11.9. Show that $\mathscr{W}_1^1$ is generated by a glide reflection and a reflection and that $\mathscr{W}_1^3$ is generated by two glide reflections with parallel axes.

11.10. Find a unit cell in each of the patterns in Figure 11.31.

11.11. Name the symmetry group for each of the wallpaper patterns in Figure 11.32.

11.12. Find a unit cell and a base for each of the patterns in Figure 11.32.

11.13. Find the symmetry group, a unit cell, and a base for each of the four wallpaper patterns in Figure 11.33.

11.14. For each of the seventeen wallpaper groups, make a page size wallpaper pattern that has that group as its symmetry group.

11.15. Prove or disprove: A wallpaper pattern with a 4-center and a line of symmetry has symmetry group $\mathscr{W}_4^1$ iff there are two lines of symmetry intersecting at an angle of 45°.

Figure 11.33

11.16. Prove or disprove: A wallpaper pattern with a 4-center and a line of symmetry has symmetry group $\mathcal{W}_4^2$ iff any two lines of symmetry are parallel or perpendicular.

11.17. Name the symmetry group of the wallpaper patterns in Figure 11.34. These are taken from W. & G. Audsley's *Designs and Patterns from Historical Ornament* in the Dover Pictorial Archive Series.

11.18. Prove or disprove: A translation lattice for a wallpaper group that is both rectangular and rhombic must have a square for a unit cell.

11.19. In Figure 11.25, find at least ten bases for $\mathcal{W}_2^3$ in the unit cell defined by $\square ABDC$.

11.20. What can you say about those groups of isometries of the plane whose subgroup of translations is generated by three translations?

Figure 11.34a

11.21. Name the symmetry group of the wallpaper patterns in Figure 11.35.

11.22. Prove: Every wallpaper group is a subgroup of either a wallpaper group $\mathcal{W}_4^1$ or of a wallpaper group $\mathcal{W}_6^1$.

11.23. Read *Fantasy and Symmetry, The Periodic Drawings of M. C. Escher* by Caroline H. Macgillavry (Abrams; New York, 1976).

11.24. For an explanation of the short international form for the wallpaper groups, read "The plane symmetry groups: their recognition and notation" by Doris Schattschneider in the *American Mathematical Monthly* **85** (1978), 439–450.

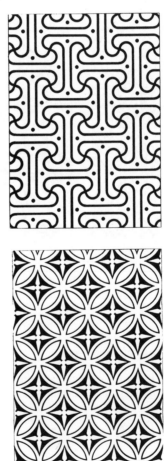

Figure 11.34b

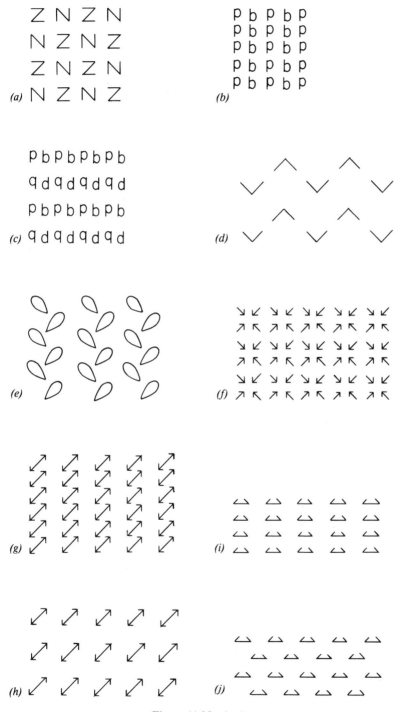

Figure 11.35 (a–j)

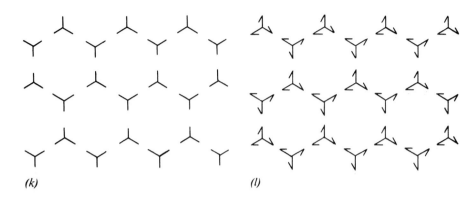

(k)   *(l)*

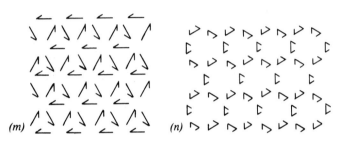

(m)   *(n)*

Figure 11.35   (k–n)

Chapter 12

Tessellations

§12.1 Tiles

Almost everyone has at one time or another been intrigued by mosaic patterns. Figure 11.31 illustrates seventeen tilings of the plane. A *tiling* of the plane is also called a *mosaic, tessellation,* or *paving* of the plane. Tiling as an art predates human history and, perhaps, reached its zenith in the Moorish forts, palaces, and mosques near the end of what Westerners call the Middle Ages. Except for an initial study by the astronomer Johannes Kepler (1571–1630), little formal mathematical investigation of tilings took place before the end of the last century. Much of what has been done is the work of chemists and crystallographers. Today, mathematicians are taking more interest in this ancient topic.

A **tessellation** or **tiling** of the plane is a set $\{T_1, T_2, \ldots\}$ of polygonal regions that cover the plane without gaps and without overlap of nonzero area. (A polygonal region T_i contains its boundary and will be called a *polygon* in this chapter. Fancier regions are possible but not necessary here.) The polygons T_i are the **tiles** of the tessellation. Every polygon is a tile for some tessellation. A tiling is respectively **monohedral, dihedral,** or **trihedral** if under congruence the tiles form one class, two classes, or three classes, respectively; in general a tiling is *r*-**hedral** if there is a set $\{P_1, P_2, \ldots, P_r\}$ of r polygons, called **prototiles**, such that each tile of the tiling is congruent to exactly one of the prototiles and if each prototile is congruent to at least one of the tiles. If P is a prototile for a monohedral tiling, we say P **admits** a tiling or P **tiles** the plane. The tiling illustrated in Figure 12.1a is dihedral with a triangle and a hexagon as its two prototiles. The tiling illustrated in Figure 12.1c is monohedral with a cross as its single prototile.

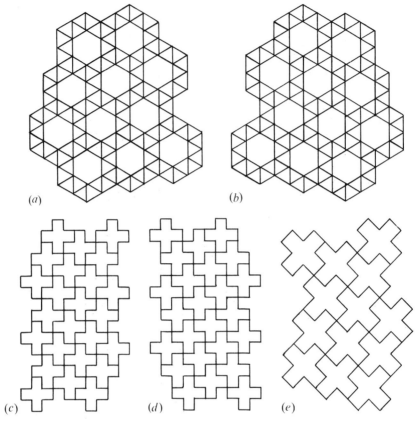

Figure 12.1

If $\{T_1, T_2, \ldots\}$ is a tiling and α is a similarity, then $\{\alpha(T_1), \alpha(T_2), \ldots\}$ is a tiling that is **similar** to the first tiling; if α is an isometry, the tilings are **congruent**. In Figure 12.1, tilings a and b are congruent, as are tilings c and d. None of these tilings has a line of symmetry. In the same figure, each of the tilings c, d, e is similar to the other. Tilings that are similar are often tacitly supposed to be "the same."

Is there anyone not familiar with the three **regular** tilings, given in Figure 12.2? Each of these monohedral tilings has a regular polygon for its prototile.

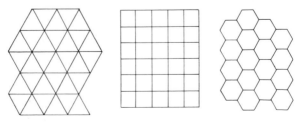

Figure 12.2

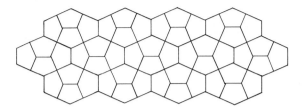

Figure 12.3

The regular pentagon does not tile the plane since three around a vertex leave a gap and four overlap. That is not to say that there are not many equilateral pentagons that tile. The beautiful *Cairo tessellation* with a convex equilateral pentagon as its prototile is illustrated in Figure 12.3. The tessellation is so named because such tiles were used for many streets in Cairo. The construction of the prototile is shown in Figure 12.4 where one starts with the base $\overline{AB}$, the midpoint of this base, and the 45° angles shown. Vertices C and E and then finally D are obtained by a compass with opening AB. The angles at C and E are right angles (Exercise 12.8). Since the sum of the measures of the angles at A, B, and D must then be 360, four copies of the prototile form the hexagon in Figure 12.4. Since this hexagon has a point of symmetry, it is seen in the next paragraph that the hexagon tiles the plane under translations alone, giving rise to the desired Cairo tessellation.

Suppose hexagon H has point L as a point of symmetry. We want to show H tiles the plane. First note that the pairs of opposite sides of a hexagon, whether convex or not, are parallel and congruent iff the hexagon has a point of symmetry. (A halfturn is a dilatation and an isometry.) Let M and N be the midpoints of sides $\overline{AB}$ and $\overline{BC}$, respectively, of hexagon H. Then $\sigma_M \sigma_L$ is a translation taking the side opposite $\overline{AB}$ onto $\overline{AB}$. The union of all images of H under all the powers of the translation $\sigma_M \sigma_L$ is a string of hexagons dividing the plane. In Figure 12.5, this string is along the left edge. The images of this string under the powers of the translation $\sigma_N \sigma_L$ cover the plane without overlap. Thus, a hexagon having a point of symmetry tiles the plane under

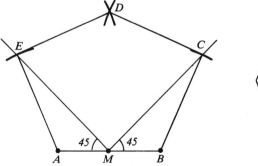
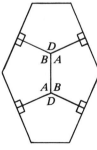

Figure 12.4

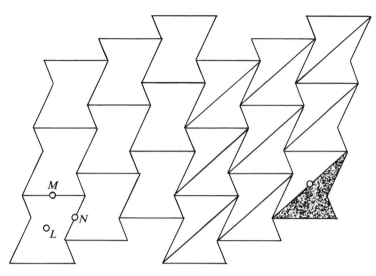

Figure 12.5

translations alone. With this little lemma, we can prove the somewhat surprising fact that every quadrilateral admits a tiling of the plane. It makes no difference whether the quadrilateral is convex or not. Let L be the midpoint of a side of any quadrilateral Q. See the lower right side of Figure 12.5. Now, quadrilaterals Q and $\sigma_L(Q)$ intersect only along their common side containing L, and their union is a hexagon H having L as a point of symmetry. Since H tiles the plane, then it is trivial that Q must tile the plane also. Therefore, every quadrilateral does tile the plane. Further, since the union of any triangle and its image under the halfturn about the midpoint of a side is a quadrilateral, it follows that any triangle also tiles the plane.

Theorem 12.1. *Any triangle tiles the plane; any quadrilateral tiles the plane. A hexagon with a point of symmetry tiles the plane.*

There is no known general procedure for telling whether or not a polygon is a prototile for a monohedral tessellation. Although, it is easy to see that the bow tie in Figure 12.6 admits a tiling, which of the other polygons in the figure tile the plane? The four polygons on the right side of Figure 12.6 are *heptominoes*. They are special cases of ***polyominoes***, which are the figures formed by

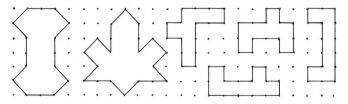

Figure 12.6

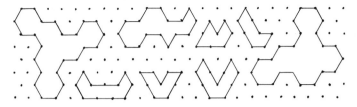

Figure 12.7

connecting unit squares edge-to-edge. Of course the union of two such rookwise connected unit squares is a *domino*. Then there are the *trominoes, tetrominoes, pentominoes, hexominoes,* etc. The polyominoes were introduced by Solomon W. Golomb, and his delightful book *Polyominoes* is a good place to start enjoying their study. ***Polyiamonds***, on the other hand, are formed by connecting congruent equilateral triangles edge-to-edge. The union of two such triangles with a common edge is a *diamond*. Then there are the *triamonds, tetriamonds, pentiamonds, hexiamonds,* etc. Polyiamonds have been stressed by Thomas H. O'Beirne in his column in the British magazine *New Scientist*. Which of the polyiamonds in Figure 12.7 are prototiles for a monohedral tessellation? The polyiamond of order 18 (second from the left on top) in Figure 12.7 is Roger Penrose's *loaded wheelbarrow*. Roger Penrose is the celebrated Oxford mathematician featured in the cover article of the December 1980 issue of *Science 80*. The loaded wheelbarrow does tile the plane. In Figure 12.7, the polyiamonds at the far right and far left are *polyhexes*. In fact, each is a *hexahex*.

Once it is shown that a particular polygon admits a tiling, there is the question, "In how many ways?" The following definition makes precise the idea that a prototile tesselates the plane in exactly k ways. If P is the only prototile for each of k mutually incongruent tilings and if any monohedral tiling having P for its prototile is congruent to one of these k tilings, then P is said to be k-***morphic***. It is not necessary to search far to find a tile that is monomorphic (i.e., 1-morphic). Dimorphic and trimorphic tiles are given in Figure 12.8. These tiles were introduced by Branko Grünbaum and G. C. Shephard. Grünbaum and Shephard have coauthored many publications on tessellations and their book *Tilings and Patterns* should be sought for

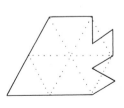 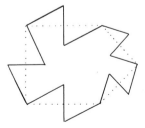

Figure 12.8

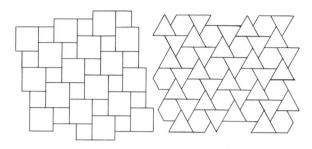

Figure 12.9

further study on this topic. The two tessellations determined by the dimorphic prototile and the three tessellations determined by the trimorphic
prototile are left for Exercise 12.18.

For the rest of this section, we restrict our attention to tessellations whose
prototiles are regular polygons. From Figure 12.9 it follows that the possibilities are infinite and we might therefore impose further restrictions. For
one thing, the polygons in Figure 12.9 do not meet edge-to-edge. A tessellation is **edge-to-edge** if each two tiles intersect along a common edge, only at a
common vertex, or not at all. It follows (Exercise 12.3) that up to similarity
the only monohedral edge-to-edge tilings by regular polygons are the three
regular tilings of Figure 12.2. The kagome tiling on the left in Figure 12.10
is dihedral and edge-to-edge. This tiling, named after a three-way bamboo
weave, has vertices surrounded by three hexagons and vertices surrounded
by two hexagons and two triangles. We would like our tilings to be edge-to-
edge and to have all vertices surrounded alike in the number of each kind of
regular polygon, in which case we say the vertices have the same **species**. That
this still leaves infinitely many possibilities can be seen from the edge-to-edge
tilings indicated on the right in Figure 12.10 and composed of alike horizontal
stripes. Here, every vertex is surrounded by two hexagons and two triangles.
This property is maintained in each of the infinitely many tessellations obtained by sliding each horizontal strip one unit to the right (or left) or else
leaving the strip in position. More interesting examples of infinite families
of edge-to-edge tilings by regular polygons where all vertices have the same

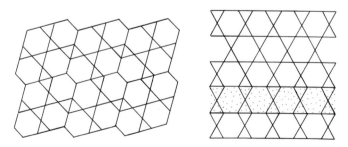

Figure 12.10

species are indicated in Exercises 12.16 and 12.17. The determination of all such tessellations is yet to be accomplished. So we would like each vertex to be surrounded not only alike in the number of each kind of regular polygon but also alike in the cyclic order of these polygons about the vertex. If a vertex is surrounded in cyclic order by regular n-gons of order $n_1, n_2, \ldots,$ and n_r, then the vertex is said to be of *type* $n_1 \cdot n_2 \cdots n_r$. Certain obvious abbreviations are also used. For example, on the right in Figure 12.10 there are vertices of type $3 \cdot 6 \cdot 3 \cdot 6$ and vertices of type $3 \cdot 3 \cdot 6 \cdot 6$. These types can be abbreviated $(3 \cdot 6)^2$ and $3^2 \cdot 6^2$, respectively. Sliding the shaded strip in the figure one unit, we then have all vertices of type $(3 \cdot 6)^2$. It turns out not to be necessary for our purpose to further subdivide the types by considering orientation (clockwise vs. counterclockwise). Hence, we want all edge-to-edge tessellations by regular polygons where all vertices have the same type.

What are the possible types for a vertex? Suppose vertex V is of type $n_1 \cdot n_2 \cdots n_r$. Since among the regular polygons the equilateral triangle has the smallest vertex angles, then $r \leq 6$, with $r = 6$ iff each n_i is 3. In general, the vertex angle of a regular n-gon has degree measure $180(n - 2)/n$. Hence, the sum of the r terms $180(n_i - 2)/n_i$ must be 360. Equivalently, we want the solutions in positive integers to the equations

$$\sum_{i=1}^{r} \frac{1}{n_i} = \frac{r - 2}{2}$$

with $3 \leq r \leq 6$. The seventeen numerical solutions are obtained by arithmetic. These solutions give the seventeen possible species. Four of these give rise to two types. Thus, there are the twenty-one types listed in Table 12.1, corresponding to the seventeen solutions.

Not all the types given by the arithmetic are possible for an edge-to-edge tessellation by regular polygons where each vertex has the same type. Suppose every vertex is of type $3 \cdot x \cdot y$. Marching around an equilateral

Table 12.1

(1) **3·3·3·3·3·3**		
(2) **3·3·3·3·6**		
(3) **3·3·3·4·4**	and	**3·3·4·3·4**
(4) 3·3·4·12	and	3·4·3·12
(5) 3·3·6·6	and	**3·6·3·6**
(6) 3·4·4·6	and	**3·4·6·4**
(7) **4·4·4·4**		
(8) 3·7·42	(9)	3·8·24
(10) 3·9·18	(11)	3·10·15
(12) **3·12·12**	(13)	4·5·20
(14) **4·6·12**	(15)	**4·8·8**
(16) 5·5·10	(17)	**6·6·6**

triangle, we see the sides of the triangle alternate as common sides with an
x-gon and a y-gon. Since 3 is odd, after one circuit we have $x = y$. This
eliminates species 8, 9, 10, and 11 in Table 12.1 but not species 12. Likewise,
the vertices cannot be of type $x \cdot 5 \cdot y$ unless $x = y$ since 5 is also odd. This
eliminates species 13 and 16. If each of A and B of equilateral $\triangle ABC$ is of
type $3 \cdot x \cdot y \cdot z$, then at vertex C the triangle lies either between two x-gons
or between two z-gons, which is impossible if A, B, C all have type $3 \cdot 3 \cdot 4 \cdot 12$,
$3 \cdot 4 \cdot 3 \cdot 12$, $3 \cdot 3 \cdot 6 \cdot 6$, or $3 \cdot 4 \cdot 4 \cdot 6$. We have eliminated all but the eleven
types in boldface in Table 12.1. Of these, only type $4 \cdot 6 \cdot 12$ has two dis-
tinguishable orientations; the other ten types read the same whether reading
clockwise or counterclockwise about a vertex. Perhaps this type will therefore
yield several tessellations. Of course, types 3^6, 4^4, and 6^3 give precisely the
three regular tessellations. Figure 12.11 can be used to demonstrate that each
of the remaining eight types does, in fact, have one realization in a desired
tiling. For each of the eight figures in Figure 12.11, pick a vertex and mark the

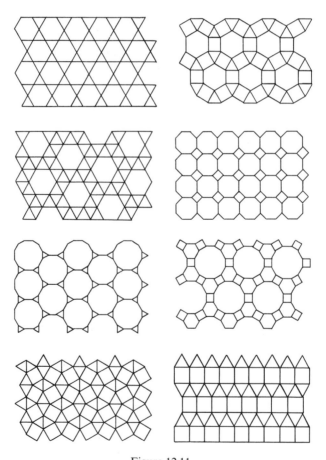

Figure 12.11

polygons sharing that vertex by placing an "n" in an n-gon. Then continue to mark those polygons whose position is forced by assuming all vertices have the same type. Several observations result from this valuable exercise. First, in the figure having a vertex of type $4 \cdot 6 \cdot 12$, it is immediately clear that both orientations of the type are required for a tiling. With both orientations allowed, the one vertex then uniquely determines the rest of the tessellation. This maverick tiling might be excluded if it did not share other nice properties with the remaining tilings. A second observation concerns the figure containing a vertex of type $3^4 \cdot 6$. Beginning with one vertex we find the rest of the tiling is not forced. At the time you have marked one hexagon and its surrounding eighteen triangles, there is a choice to be made concerning the placing of the next hexagon. Making one of the two possible choices for placing the next hexagon on one side of the hexagonal figure consisting of the first hexagon and its surrounding triangles, the remaining part of the tiling is then forced. The two tilings obtained in this manner are congruent, however. One is the image of the other under a reflection. (The two forms are said to be *enantiomorphic*; in general, any figure without a line of symmetry and its image under an odd isometry are *enantiomorphs* of each other.) See Figure 12.1 again. This is the only tiling of the eight without a line of symmetry. For another peculiarity of this tiling see Exercise 12.1. No problem is encountered in the unique extension of the remaining types to a tiling. Hence, we have the surprising result that, up to similarity, each type gives rise to a unique tessellation of the plane. Therefore, the tiling itself can be unambiguously named after the type of its vertices. These eight tessellations are said to be the **semiregular** or **Archimedean** tessellations of the plane. Since Kepler is the first person known to exhibit the regular and semiregular tilings, it seems only fair to call the result of our search **Kepler's Theorem**.

Theorem 12.2. *Up to similarity, there are exactly eleven edge-to-edge tessellations whose prototiles are regular polygons and such that all vertices have the same type.*

A tessellation is said to be **vertex transitive, edge transitive**, or **tile transitive** if given two vertices, edges, or tiles, respectively, there is a symmetry of the tessellation that takes one to the other. Another surprising property shared by the Archimedean tilings is that they are all vertex transitive. So not only the immediate neighborhood looks the same from each vertex, but the whole tiling looks the same from each vertex. Only one of the Archimedean tessellations is edge transitive. Which? The two tilings in Figure 12.12 are both vertex transitive and both edge transitive, and one of them is also tile transitive. The three regular tessellations are tile transitive, but none of the Archimedean tessellations could be tile transitive. The tiling on the right in Figure 12.12 has the property that for any two given congruent tiles of the tessellation there is a symmetry of the tessellation that takes one of these tiles to the other. Is this property shared by the Archimedean tessellations?

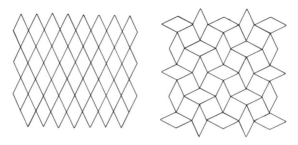

Figure 12.12

§12.2 Reptiles

Which convex polygons admit a monohedral tiling of the plane? The answer
to this question is unknown, again. The "again" will take some explaining.
It can be shown that no tiling has as its set of prototiles a finite set of convex
polygons each having more than six sides. A reference for this result is given
in Exercise 12.9. So any monohedral tiling by a convex polygon must have a
prototile that is a hexagon, a pentagon, a quadrilateral, or else a triangle. It
has long been known that a convex hexagon tiles the plane iff vertices A, B, C,
D, E, F can be named in cyclic order such that one of the following conditions
is satisfied:

(1) $m\angle A + m\angle B + m\angle C = 360, CD = FA.$
(2) $m\angle A + m\angle B + m\angle D = 360, BC = DE, CD = FA.$
(3) $m\angle A = m\angle C = m\angle E = 120, AB = BC, CD = DE, EF = FA.$

We know every quadrilateral tiles the plane and every triangle tiles the plane.
Thus, there remains only the question of which convex pentagons tile the
plane. The problem was presumed to be solved in 1918 and independently
again in 1963. However, in 1968, R. B. Kershner announced in the *American
Mathematical Monthly* that in addition to the previously known five types
of convex pentagons that admit a tiling there were three new types. (A
property of these new types will be discussed below.) Martin Gardner described what was thought to be the complete list of tiling convex pentagons
in his celebrated column "Mathematical Games" from the magazine
Scientific American (July 1975). As a puzzle, Richard E. James III thought he
would try to determine the list before he finished reading the article. The
result of this was Gardner's announcement in his December column that
James had found a family overlooked by Kershner. Marjorie Rice, a mathematician with no mathematical training beyond high school, has since
found several additional families. Whether the present list is complete is not
known. Even the list of all convex pentagons admitting an edge-to-edge
tiling and the list of all equilateral convex pentagons admitting a tiling are
unknown.

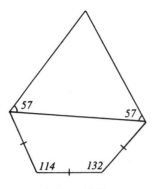

Figure 12.13

David Hilbert (1862–1943), who is frequently called the leading mathematician of this century, addressed the second International Congress of Mathematicians at Paris in 1900. Here he proposed a list of problems which he thought should occupy the attention of mathematicians in the twentieth century. Indeed, these problems have greatly influenced the direction of mathematical research since 1900. Some of the problems had several parts and some were not well formulated. One question had to do with tiling. You should be able to think of a monohedral tiling by dominoes that has the identity as its only symmetry. However it is even easier to find a tiling by dominoes that is tile transitive. Is it true that a prototile for any monohedral tiling also admits a tiling that is tile transitive? Hilbert apparently thought so. He was proved wrong in 1935 with a counterexample having a nonconvex prototile. The new families of pentagons discovered by Kershner provided examples of convex polygons admitting only edge-to-edge tilings that are not tile transitive. The construction of one such pentagon is given in Figure 12.13.

For each frieze group, Exercise 12.5 asks for a monohedral tiling by dominoes that has the frieze group as its symmetry group. These tilings are invariant under translations in only one independent direction. However, there certainly is a tiling by dominoes that has a wallpaper group for its symmetry group. A tiling whose symmetry group is neither a wallpaper group nor a frieze group must have a finite symmetry group C_n or D_n. Such a tiling is said to be *nonperiodic*. The monohedral tiling in Figure 12.14 is nonperiodic, with symmetry group C_2. However, the prototile of this tiling does admit a tiling with a wallpaper group as its symmetry group (Exercise 12.11). One of the major problems in tiling theory is the following. Does there exist a prototile admitting only nonperiodic tilings? The problem can be stated in a more severe form. Is there a monomorphic prototile whose monohedral tiling has symmetry group C_1? Roger Penrose, mentioned in the previous section, has invented a pair of tiles that give only dihedral tilings that are nonperiodic. The two prototiles are constructed from a rhombus, as in Figure 12.15 where the measure of each angle is a multiple of 36, but instead of adding bumps and dents to give the tiles the desired matching properties,

Figure 12.14

the tiles are colored as in the figure. When placed next to each other in a tiling, the circles around a vertex must be of one color. Thus, the position in Figure 12.15 is not allowed in a tiling. The tiles are called *kites* and *darts*. The reader is warned that tiling with the Penrose kites and darts can be habit forming. Replacing the colored circles by your own bumps and dents to force only allowed juxtapositions can be fun. Penrose himself has changed the two prototiles into two chickens. These may be seen in Martin Gardner's column in the January 1977 *Scientific American.* Penrose's pair of nonperiodic chickens comprise an advanced piece of mathematics.

Solomon W. Golomb, who introduced polyominoes, is also responsible for focusing attention on replicating figures in the plane. If polygon P can be cut into k congruent polygons (with shared boundaries) each of which is similar to polygon Q, then we say Q *divides* P with *multiplicity* k. If polygon P divides itself with multiplicity greater than 1, then we follow Golomb and say P is a *reptile* (or replicating figure). To say that polygon P is a reptile that divides itself with multiplicity k, we simply say P is *rep-k*. The hexiamond outlining Figure 12.16 is called the *sphinx* and is evidently rep-4. In fact, the sphinx is the only known rep-4 pentagon. Can you show the sphinx is also rep-9? Examples of rep-4 hexagons are given in Figure 12.17. The tromino

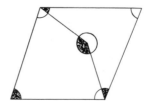

Figure 12.15

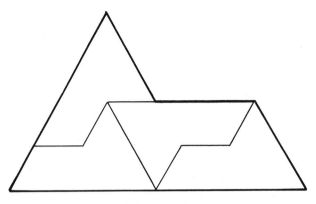

Figure 12.16

in the figure is a special case of the reptile formed by removing a quadrant from any rectangle. The pentomino on the right in the figure and the tromino are also both rep-9. Rather than view reptiles as dissection problems, we notice that polygon P is rep-k iff k congruent copies of P can be assembled without overlap (except along boundaries) into a polygon Q that is similar to P. Then, of course, k copies of Q can be assembled into a still larger polygon similar to all the others. Each of the k copies of Q contains k copies of P. Continuing in this fashion, we obtain larger and larger polygons each similar to P and each divided into polygons congruent to P. We claim it follows that P tiles the plane. This appears obvious only at first. Some of the difficulties become apparent when you try to verify the result for a particular reptile, say the sphinx. For a general proof, see the Extension Theorem in the book *Tilings and Patterns* by Grünbaum and Shephard, mentioned earlier. For any of the individual reptiles we shall encounter, these difficulties can be overcome, and we shall suppose that any reptile tiles the plane.

For any positive integer k, there is a rep-k reptile. Any 1 by $\sqrt{k}$ parallelogram is rep-k since $1/\sqrt{k} = \sqrt{k}/k$. Parallelograms are the only known rep-7 reptiles. Figure 12.18 illustrates a rep-3 parallelogram and the only three known examples of rep-4 quadrilaterals that are not parallelograms.

The altitude to the hypotenuse of an isosceles right triangle divides the triangle into two congruent triangles each similar to the original triangle. Hence an isosceles right triangle is rep-2. It is not too difficult to prove that the only rep-2 triangles are the right isosceles triangles, that the only rep-2

Figure 12.17

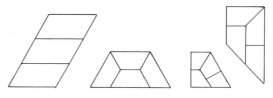

Figure 12.18

quadrilaterals are the 1 by $\sqrt{2}$ parallelograms, and that any convex rep-2 reptile is an isosceles right triangle or else a 1 by $\sqrt{2}$ parallelogram. A 30–60–90 triangle is easily seen to be rep-3.

The parallelogram in the middle of Figure 12.19 is rep-3^2, as are the two triangles determined by a diagonal of the parallelogram. From this special case, we generalize to show first that given a parallelogram P and a positive integer k, the parallelogram is rep-k^2. Consider the points on the diagonal of P that divide the diagonal into k congruent segments. The lines through these points and parallel to the sides of P then determine the k^2 congruent parallelograms similar to P. So P is rep-k^2. Using a halfturn about the midpoint of a side of a triangle, we see every triangle is half a parallelogram. It quickly follows that, given a triangle and a positive integer k, the triangle is rep-k^2.

Suppose positive integer k is given. Since each of the three smaller triangles in the 30–60–90 rep-3 triangle in Figure 12.19 is itself rep-k^2, then the 30–60–90 triangle is rep-$3k^2$. That there also exists a rep-$2k^2$ triangle is a special case of the argument that if $n = a^2 + b^2$ then there is a rep-n triangle. We suppose we are given positive integers a and b. See Figure 12.19 again. The altitude to the hypotenuse divides a right triangle T with legs of lengths a and b into two smaller triangles A and B, each similar to T. Now A, the smaller triangle with hypotenuse of length a, is rep-a^2, and B, the smaller triangle with hypotenuse of length b, is rep-b^2. So we have $a^2 + b^2$ triangles each similar to T. However, are the a^2 triangles in A and the b^2 triangles in B congruent?

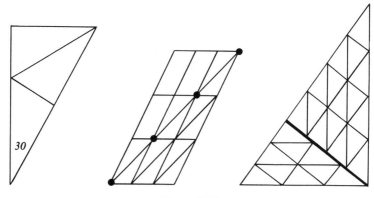

Figure 12.19

Yes, since the ratio of the area of A to the area of B is a^2/b^2, then the $a^2 + b^2$ triangles all have the same area. Similar triangles with the same area are congruent. Hence T is rep-n where $n = a^2 + b^2$. (Students of the theory of numbers will recall that $n = a^2 + b^2$ iff the square free part of n is not divisible by a prime of the form $4t - 1$.)

Theorem 12.3. *Every parallelogram is rep-k^2 and every triangle is rep-k^2, for any positive integer k. If $n = a^2 + b^2$ or $n = 3a^2$ where a and b are integers with $a \neq 0$, then there is a rep-n triangle. For any positive integer k, there is a rep-k parallelogram.*

The following follows directly from the definition of multiplicity.

Theorem 12.4. *Let P, Q, R be polygons. If P divides Q with multiplicity s and Q divides P with multiplicity r, then P is rep-rs. If P divides Q with multiplicity s, Q divides R with multiplicity t, and R divides P with multiplicity u, then P is rep-stu.*

The hexiamonds have rather fancy names in the mathematical literature. Hexiamonds *lobster, snake, butterfly,* and *bat* are illustrated in Figure 12.20. A snake cannot be a reptile since a snake cannot fit into a 60° angle and contain the vertex of that angle. However, a lobster divides a 2 by 3 parallelogram with multiplicity 2, and a 2 by 3 parallelogram divides a lobster with multiplicity 18. Hence, by the first part of Theorem 12.4, a lobster is rep-36.

Theorem 12.5. *A lobster is a reptile; a snake is not a reptile.*

The second part of Theorem 12.4 is illustrated by the following. The *stairs* and F hexominoes in Figure 12.21 divide a 3 by 4 rectangle with multiplicity 2; a 3 by 4 rectangle divides a square with multiplicity 12; and a square divides a hexomino with multiplicity 6. Hence, these two hexominoes are rep-144. Hexomino J in Figure 12.21 also divides a 3 by 4 rectangle with multiplicity 2;

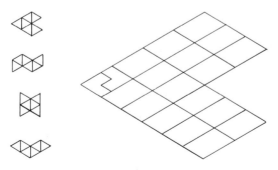

Figure 12.20

Figure 12.21

a 3 by 4 rectangle divides a 1 by 2 rectangle with multiplicity 6; and a 1 by 2 rectangle divides the J with multiplicity 3. Hence, the J hexomino is rep-36.

Finally, we mention some ways of generating reptiles. From the center of a $2n$ by $2n$ checker board to an edge, trace a path along boundaries of the colored squares such that, except at the center, the path does not intersect its other images under successive rotations of $90°$ about the center. The images define a n^2-omino that is rep-$4n^2$. See Figure 12.22. Similar constructions beginning with $3m$ triangles on the side of an equilateral triangle give rise to $3m^2$-iamonds that are rep-$9m^2$.

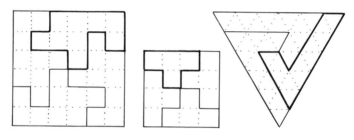

Figure 12.22

§12.3 Exercises

12.1. For which Archimedean tilings is it true that given two congruent tiles there is a symmetry of the tessellation taking one tile onto the other. Which Archimedean tilings are edge transitive?

12.2. Cut out of heavy paper a convex quadrilateral templet having angles of measure (approximately) 50, 70, 110, and 130. By tracing around the templet, draw a monohedral tiling of the plane having the quadrilateral as its prototile.

12.3. Show any monohedral edge-to-edge tiling by a regular polygon is one of the three regular tessellations.

12.4. The *dual* of an Archimedean tiling is obtained by joining the centers of adjacent tiles. Figure 12.23 shows $3^2 \cdot 4 \cdot 3 \cdot 4$ and its dual. Draw the dual for the other Archimedean tilings.

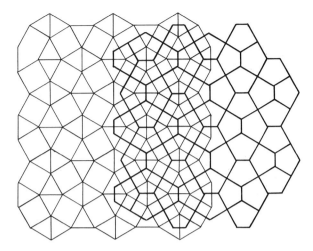

Figure 12.23

12.5. For each of the seven frieze groups and for each of the groups C_1, C_2, C_4, D_1, and D_2, give a monohedral tiling by dominoes that has this group as its symmetry group.

12.6. Construct a rep-12 triangle, a rep-16 triangle, a rep-16 quadrilateral, and a rep-24 quadrilateral.

12.7. True or False
   (a) Every trapezoid admits a tiling of the plane.
   (b) If polygon P divides polygon Q and polygon Q divides P, then both P and Q are reptiles.
   (c) Suppose P, Q, R are polygons. If P divides Q with multiplicity a, Q divides R with multiplicity b, and R divides P with multiplicity c, then each of P, Q, R is rep-abc.
   (d) Once one tile is placed, the position of all the remaining tiles of a tessellation by a monomorphic prototile is determined.
   (e) Any pentagon having a pair of parallel sides admits a tiling.
   (f) With x and y positive rational numbers, an x by y rectangle divides a square.
   (g) Given square S and rectangle R, then S divides R.
   (h) A polygon that divides a square is a reptile.
   (i) A tiling whose prototiles are all regular polygons is called an Archimedean tiling.
   (j) If k is the product of positive integers m and n, then a $\sqrt{m}$ by $\sqrt{n}$ parallelogram is rep-k.

12.8. Show the prototile constructed in Figure 12.4 for the Cairo tessellation has two right angles.

12.9. Read "Convex pologons that cannot tile the plane" by Ivan Nivin in the December 1978 (volume 85) *American Mathematical Monthly* (pp. 785–792).

12.10. Show the sphinx and each of the polygons in Figure 12.18 are rep-9.

12.11. Show the prototile of the monohedral tessellation in Figure 12.14 tiles the plane in a pattern having a wallpaper group as its symmetry group.

12.12. Which polygons in Figure 12.6 admit a tiling?

12.13. Show that the only rep-2 triangles are the isosceles right triangles, that the only rep-2 quadrilaterals are the 1 by $\sqrt{2}$ parallelograms, and that any convex rep-2 reptile is an isosceles right triangle or a 1 by $\sqrt{2}$ parallelogram.

12.14. Show Penrose's loaded wheelbarrow admits a tiling. Which other polyiamonds in Figure 12.7 admit a tiling?

12.15. Prove the sphinx and the middle polygon in Figure 12.17 each tile the plane.

12.16. Show there are infinitely many edge-to-edge tilings by equilateral triangles and squares such that all vertices have the same species.

12.17. Show there are infinitely many edge-to-edge tilings by regular polygons such that each vertex has species 6 from Table 12.1.

12.18. Illustrate the two tessellations by the dimorphic prototile and three tessellations by the trimorphic prototile in Figure 12.8.

12.19. On a sheet of graph paper, join $(0, 0)$ to $(0, 1)$ to $(4, 3)$ to $(4, 4)$ and join $(0, 4)$ to $(1, 4)$ to $(3, 0)$ to $(4, 0)$. To these points add their images under the reflection in the X-axis, the reflection in the Y-axis, and the halfturn about $(0, 0)$. Consider this augmented set together with its images under all translations with equations $x' = x + 8h$ and $y' = y + 8k$ where h and k are integers. Describe what you have and name the symmetry group for the set.

12.20. Show that if D is in the interior of $\triangle ABC$, point M is the midpoint of $\overline{AC}$, and $E = \sigma_M(D)$, then A, B, C, D, E are the vertices of a pentagon that tiles the plane.

12.21. Which of the hexominoes in Figure 12.24 are reptiles?

Figure 12.24

12.22. The polyomino with vertices $(0, 0), (1, 0), (1, 4), (0, 4), (0, 3), (-1, 3), (-1, 2)$, and $(0, 2)$ is a Y pentomino. Is this polygon a reptile?

12.23. Draw a picture of each of the twelve hexiamonds: bar, butterfly, bat (chevron), crown, hexagon, crook (club), hook (shoe), lobster, pistol (signpost), snake, sphinx, and yacht. Which hexiamonds are reptiles?

12.24. The F octomino is formed by rows of 3, 1, 2, 1, 1 squares. How many tilings does this octomino admit?

12.25. Show Kershner's tile in Figure 12.13 admits only edge-to-edge tilings that are not tile transitive.

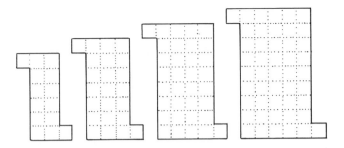

Figure 12.25

12.26. Prove or disprove: Figure 12.25 defines an infinite sequence of dimorphic prototiles for monohedral tilings. The prototiles are polyominoes of order $t^2 - 2$ for
$t = 4, 5, 6, \ldots$.

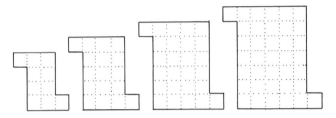

Figure 12.26

12.27. Prove or disprove: Figure 12.26 defines an infinite sequence of trimorphic prototiles for monohedral tilings. The prototiles are polyominoes of order $s^2 + 1$ for
$s = 3, 4, 5, \ldots$.

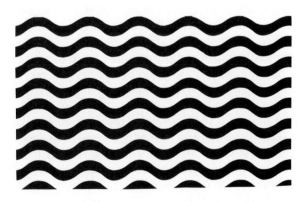

Chapter 13

Similarities on the Plane

§13.1 Classification of Similarities

This chapter may be read following Chapter 9.

The image of a triangle as seen through a magnifying glass is similar to the original triangle. We shall see that any two similar triangles in the plane are related by such a magnification and an isometry. The transformation that sends (x, y) to $(2x, 2y)$ is a magnifying glass for the Cartesian plane, multiplying all distances by 2. We shall call this mapping a *stretch*. The inverse transformation, which sends (x, y) to $(x/2, y/2)$, will not be called a *shrink* here but will also be called a *stretch*. Be warned, the language of similarity theory is *not* standardized! In general, it is impossible to tell what any of the following words means without examining the context in which that word is used: stretch, dilation, dilatation, homothety, enlargement, contraction, central similarity, radial transformation, or size transformation. (Note that di-lation and dil-a-tation are different words.) For easy reference, the next paragraph contains all of the definitions of the new transformations needed for our study of similarities.

If C is a point and $r > 0$, then a ***stretch of ratio*** r ***about*** C is the transformation that fixes C and otherwise sends point P to point P' where P' is the unique point on $\overrightarrow{CP}$ such that $CP' = rCP$. We allow the identity to be a stretch. A ***dilation about point*** C is a stretch about C or else a stretch about C followed by the halfturn about C. A ***stretch reflection*** is a nonidentity stretch about some point C followed by the reflection in some line through C. A ***stretch rotation*** is a nonidentity stretch about some point C followed by a nonidentity rotation about C. If $r > 0$, then a ***similarity of ratio*** r is a transformation α such that $P'Q' = rPQ$ for all points P and Q where $P' = \alpha(P)$ and $Q' = \alpha(Q)$.

Before going any further, reread the preceding paragraph at least twice to get the definitions firmly in mind. Recall that a dilatation is a collineation α such that $l \parallel \alpha(l)$ for every line l and that the group $\mathcal{H}$ generated by the half-turns is contained in the group $\mathcal{D}$ of all dilatations. Since a similarity is a transformation that multiplies all distances by some positive number, then the image of a triangle under a similarity is a triangle. It follows that collinear points are mapped onto collinear points by a similarity transformation, since the inverse of a similarity is also a similarity. Thus, a similarity is a collineation. The remainder of the first theorem should be evident from the definitions. Group $\mathcal{S}$ is always the group of all similarities.

Theorem 13.1. *An isometry is a similarity. A similarity with two fixed points is an isometry. A similarity with three noncollinear fixed points is the identity. A similarity is a collineation that preserves betweenness, midpoints, segments, rays, triangles, angles, angle measure, and perpendicularity. The composite of a similarity of ratio r and a similarity of ratio s is a similarity of ratio rs. The similarities form a group $\mathcal{S}$ that contains the group $\mathcal{I}$ of all isometries.*

To show a dilation is a dilatation and a similarity, first suppose α is a stretch of ratio r about point C. Transformation α fixes the lines through C. Suppose P, Q, R are three collinear points on a line off C and have images P', Q', R', respectively, under α. See Figure 13.1. Since $CP' = rCP$, $CQ' = rCQ$, and $CR' = rCR$, then it follows from the theory of similar triangles that $\overleftrightarrow{P'Q'} \parallel \overleftrightarrow{PQ}$, that points P', Q', R' are collinear, and that $P'Q' = rPQ$. Hence, a stretch is a dilatation and a similarity. Since a halfturn is a dilatation and a similarity, then the composite of a stretch and a halfturn is both a dilatation and a similarity.

Theorem 13.2. *A dilation is a dilatation and a similarity.*

Suppose $\overleftrightarrow{AB} \parallel \overleftrightarrow{A'B'}$ and there is a dilatation δ such that $\delta(A) = A'$ and $\delta(B) = B'$. See Figure 13.2. If point P is off $\overleftrightarrow{AB}$, then $\delta(P)$ is uniquely determined as the intersection of the line through A' that is parallel to $\overleftrightarrow{AP}$ and the line through B' that is parallel to $\overleftrightarrow{BP}$. Then, if Q is on $\overleftrightarrow{AB}$, point $\delta(Q)$ is

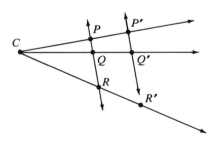

Figure 13.1

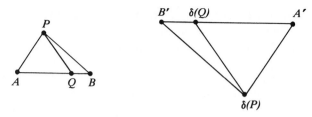

Figure 13.2

uniquely determined as the intersection of $\overleftrightarrow{A'B'}$ and the line through $\delta(P)$ that is parallel to $\overleftrightarrow{PQ}$. Since the image of each point is uniquely determined by the images of A and B, then there is at most one dilatation δ taking A to A' and B to B'. On the other hand, $\tau_{A, A'}$ followed by the dilation about A' that takes $\tau_{A, A'}(B)$ to B' is a dilatation taking A to A' and B to B'.

Theorem 13.3. *If* $\overleftrightarrow{AB} \parallel \overleftrightarrow{A'B'}$, *then there is a unique dilatation* δ *such that* $\delta(A) =$ A' *and* $\delta(B) = B'$.

If dilatation δ does not fix point A and if $A' = \delta(A)$, then $\delta(\overleftrightarrow{AA'})$ must be the line through $\delta(A)$ that is parallel to $\overleftrightarrow{AA'}$. This simple observation is our next theorem.

Theorem 13.4. *If point* A *is not fixed by dilatation* δ, *then line* $\overleftrightarrow{AA'}$ *is fixed by* δ *where* $A' = \delta(A)$.

We can now answer the question, "What are the dilatations?" A non-identity dilatation α must have some nonfixed line l. So l and $\alpha(l)$ are distinct parallel lines. Any two points A and B on line l are such that neither $\alpha(A)$ nor $\alpha(B)$ is on l. Let $A' = \alpha(A)$ and $B' = \alpha(B)$. See Figure 13.3. Now, $\overleftrightarrow{AB}$ and $\overleftrightarrow{A'B'}$ are distinct parallel lines. If $\overleftrightarrow{AA'} \parallel \overleftrightarrow{BB'}$, then $\square AA'B'B$ is a parallelogram, $\tau_{A, A'}(B) = B'$, and (Theorem 13.3) dilatation α must be the translation $\tau_{A, A'}$. However, suppose $\overleftrightarrow{AA'} \nparallel \overleftrightarrow{BB'}$. Then the lines $\overleftrightarrow{AA'}$ and $\overleftrightarrow{BB'}$ are fixed (Theorem 13.4) and must intersect at some fixed point C. Since $\overleftrightarrow{AB}$ is not fixed, then C is off both parallel lines $\overleftrightarrow{AB}$ and $\overleftrightarrow{A'B'}$ with C, A', A collinear and C, B', B

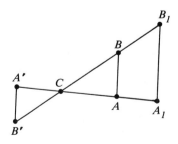

Figure 13.3

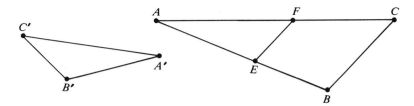

Figure 13.4

collinear. So $CA'/CA = CB'/CB$. Then there is a dilation δ about C such that $\delta(A) = A'$ and $\delta(B) = B'$. (If point C is between points A and A', then δ is a stretch followed by σ_C; otherwise, δ is simply a stretch about C.) By the uniqueness of a dilatation taking A to A' and B to B', the dilatation α must be the translation $\tau_{A,\,A'}$ or else the dilation δ.

Theorem 13.5. *A dilatation is a translation or a dilation.*

To show there is a similarity taking one triangle onto any similar triangle, suppose $\triangle ABC \sim \triangle A'B'C'$. See Figure 13.4. Let δ be the stretch about A such that $\delta(B) = E$ with $AE = A'B'$. With $F = \delta(C)$, then $\triangle AEF \cong \triangle A'B'C'$ by ASA. Since there is an isometry β such that $\beta(A) = A'$, $\beta(E) = B'$, and $\beta(F) = C'$, then $\beta\delta$ is a similarity taking A, B, C to A', B', C', respectively. If α is a similarity taking A, B, C to A', B', C', respectively, then $\alpha^{-1}(\beta\delta)$ fixes three noncollinear points and must be the identity. Therefore, $\alpha = \beta\delta$, and we have the following analogue to Theorem 5.7.

Theorem 13.6. *If $\triangle ABC \sim \triangle A'B'C'$, then there is a unique similarity α such that $\alpha(A) = A'$, $\alpha(B) = B'$, and $\alpha(C) = C'$.*

Generalizing from triangles to arbitrary sets of points, we say figures s and t are ***similar*** if there is a similarity α such that $\alpha(s) = t$.

By the proof above, a similarity is just a stretch about some point P followed by an isometry. Actually, the point P can be arbitrarily chosen as follows. If α is a similarity of ratio r, let δ be the stretch of ratio r about P. Then δ^{-1} is a stretch of ratio $1/r$. So $\alpha\delta^{-1}$ is an isometry and $\alpha = (\alpha\delta^{-1})(\delta)$.

Theorem 13.7. *If α is a similarity and P is any point, then $\alpha = \beta\delta$ where δ is a stretch about P and β is an isometry.*

This important theorem gives us a feeling for the nature of the similarities. We need only one more theorem on similarities before the classification theorem. However, the proof uses a lemma about directed distance, which must be introduced next. Directed distance will play a major role in the next chapter.

We suppose the lines in the plane are directed (in an arbitrary fashion) and $\underline{AB}$ denotes the directed distance from A to B on line $\overleftrightarrow{AB}$. For any points A and B, we have $\underline{AB} = -\underline{BA}$ and so $\underline{AA} = 0$. Note that for distinct points A, B, C, on line l the number $\underline{AC}/\underline{CB}$ is independent of the choice of positive direction on line l, as changing the positive direction would change the sign of both numerator and denominator and leave the value of the fraction itself unchanged. The lemma we wish to prove is the following.

Theorem 13.8. *If* $y \neq -1$, *then there exists a unique point* P *on* $\overleftrightarrow{AB}$ *different from* B *such that* $\underline{AP}/\underline{PB} = y$. *If* P *is a point on* $\overleftrightarrow{AB}$ *different from* B, *then* $\underline{AP}/\underline{PB} \neq -1$ *and* P *is between* A *and* B *iff* $\underline{AP}/\underline{PB}$ *is positive.*

To begin the proof, observe that there is a one-to-one correspondence between the points X on $\overleftrightarrow{AB}$ and real numbers x given by the equation $\underline{AX} = x(\underline{AB})$. Then $\underline{XB} = \underline{XA} + \underline{AB} = \underline{AB} - \underline{AX} = (1 - x)(\underline{AB})$. Let $f(x) = x/(1 - x)$. So $f(x) = \underline{AX}/\underline{XB}$ and $f(x) \neq -1$. Also, as $f(x) = f(z)$ implies $x = z$ for numbers x and z so $\underline{AX}/\underline{XB} = \underline{AZ}/\underline{ZB}$ implies $X = Z$ for points X and Z on $\overleftrightarrow{AB}$. In other words, point P in the first statement of Theorem 13.8 is unique provided P exists. For the existence of P, if $y \neq -1$, let P be the point such that $\underline{AP} = [y/(1 + y)](\underline{AB})$. Then $\underline{AP}/\underline{PB} = y$, as desired. The last part of the theorem then depends only on the fact that point P on $\overleftrightarrow{AB}$ is between A and B iff nonzero numbers $\underline{AP}$ and $\underline{PB}$ have the same sign.

The lemma above will now be used to show a similarity that is not an isometry must have a fixed point. Suppose α is a similarity that is not an isometry. We may suppose α is not a dilatation. (Why?) So there is a line l such that $l' \nparallel l$ where $l' = \alpha(l)$. Let l intersect l' at point A. With $A' = \alpha(A)$, then A' is on l'. See Figure 13.5. We suppose $A' \neq A$. (Why?) Let m be the line through A' that is parallel to l. With $m' = \alpha(m)$, then $m' \parallel l'$. Let m' intersect m at point B. With $B' = \alpha(B)$, then B' is on m' and distinct from A'. We suppose $B' \neq B$. So $l' = \overleftrightarrow{AA'}$, $m' = \overleftrightarrow{BB'}$, and $\overleftrightarrow{AA'} \parallel \overleftrightarrow{BB'}$. Now $\overleftrightarrow{AB} \nparallel \overleftrightarrow{A'B'}$ as otherwise $A'B' = AB$ and α is an isometry. So $\overleftrightarrow{AB}$ and $\overleftrightarrow{A'B'}$ intersect at some point P off both parallel lines $\overleftrightarrow{AA'}$ and $\overleftrightarrow{BB'}$ with P, A, B collinear and P, A', B' collinear. So $\underline{AP}/\underline{PB} = \underline{A'P}/\underline{PB'}$. If α has ratio r and $P' = \alpha(P)$, then $\underline{AP}/\underline{PB} = r\underline{AP}/r\underline{PB} = \underline{A'P'}/\underline{P'B'}$. Hence, $\underline{A'P}/\underline{PB'} = \underline{A'P'}/\underline{P'B'}$. Point P is between A'

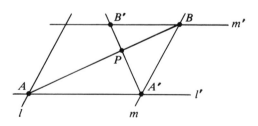

Figure 13.5

and B' iff P is between A and B since $\overleftrightarrow{AA'} \parallel \overleftrightarrow{BB'}$, but P is between A and B iff P' is between A' and B'. Hence P is between A' and B' iff P' is between A' and B'. Therefore, by the lemma above, $\underline{A'P/PB'} = \underline{A'P'/P'B'}$ and $P = P'$. So $\alpha(P) = P$, as desired.

Theorem 13.9. *A similarity without a fixed point is an isometry.*

In order to classify the similarities, suppose α is a similarity that is not an isometry. Then α has some fixed point C. So $\alpha = \beta\delta$ where δ is a stretch about C and where β is an isometry. Since $\beta(C) = \alpha\delta^{-1}(C) = C$, then β must be one of the identity ι, a rotation ρ about C, or a reflection σ_c with C on c. Hence, α is one of δ, $\rho\delta$, or $\sigma_c\delta$. We have proved the major part of *The Classification Theorem for Similarities on the Plane*.

Theorem 13.10. *A nonidentity similarity is exactly one of the following*:

$$isometry, \quad stretch, \quad stretch\ rotation, \quad stretch\ reflection.$$

There remains only the task of verifying the "exactly" in the statement of the classification theorem. This is left for Exercise 13.1.

§13.2 Equations for Similarities

There are several theorems concerning the conjugate of one similarity by another. In proving some of these, we suppose everywhere below that α is a similarity of ratio r. If γ is an isometry, then the conjugate $\alpha\gamma\alpha^{-1}$ of γ by α has ratio $(r)(1)(1/r)$ and must also be an isometry. Now suppose δ is a dilatation. Then $\alpha^{-1}(l) \parallel \delta\alpha^{-1}(l)$ for any line l by the definition of a dilatation. Since α is a collineation and $\alpha\alpha^{-1}(l) = l$, then $l \parallel \alpha\delta\alpha^{-1}(l)$ for any line l. So $\alpha\delta\alpha^{-1}$ must also be a dilatation. If η is in the group generated by the half-turns, then so is $\alpha\eta\alpha^{-1}$ since $\alpha\eta\alpha^{-1}$ is both an isometry and a dilatation by the results above, $(\mathscr{H} = \mathscr{D} \cap \mathscr{I})$. If τ is a translation, then so is $\alpha\tau\alpha^{-1}$ since $\alpha\tau\alpha^{-1}$ is in the group generated by the halfturns but is not an involution. On the other hand, $\alpha\sigma_P\alpha^{-1}$ is in the group generated by the halfturns and is an involution fixing $\alpha(P)$. So $\alpha\sigma_P\alpha^{-1}$ must be the halfturn about $\alpha(P)$. Finally, $\alpha\sigma_l\alpha^{-1}$ is a nonidentity isometry fixing every point on line $\alpha(l)$ and must be the reflection in $\alpha(l)$. All these results are put together in the following.

Theorem 13.11. *Suppose $\alpha \in \mathscr{S}$. Then*

$$\alpha\gamma\alpha^{-1} \in \mathscr{I} \quad if\ \gamma \in \mathscr{I}, \qquad \alpha\delta\alpha^{-1} \in \mathscr{D} \quad if\ \delta \in \mathscr{D},$$

$$\alpha\eta\alpha^{-1} \in \mathscr{H} \quad if\ \eta \in \mathscr{H}, \qquad \alpha\tau\alpha^{-1} \in \mathscr{T} \quad if\ \tau \in \mathscr{T},$$

$$\alpha\sigma_P\alpha^{-1} = \sigma_{\alpha(P)}, and \qquad \alpha\sigma_l\alpha^{-1} = \sigma_{\alpha(l)}.$$

In order to look at the dilatations a little more closely, a notation for the dilations is introduced as follows. If $x > 0$, then $\delta_{P,x}$ is the stretch about P of (similarity) ratio x and dilation $\delta_{P,-x}$ is defined by $\delta_{P,-x} = \sigma_P \delta_{P,x}$. Multiplying both sides of this last equation by σ_P on the left, we have $\sigma_P \delta_{P,-x} = \delta_{P,x}$. So $\delta_{P,-r} = \sigma_P \delta_{P,r}$ for all $r \neq 0$. The number r is called the **dilation ratio** of dilation $\delta_{P,r}$. There are the two special cases where a dilation is also an isometry: $\delta_{P,1} = \iota$ and $\delta_{P,-1} = \sigma_P$. Clearly, the ratio of $\delta_{P,r}$ is the absolute value $|r|$ of the dilation ratio r. For example, $\delta_{P,-3}$ has ratio $+3$ but dilation ratio -3. Since $\sigma_P \delta_{P,r} = \delta_{P,r} \sigma_P$ follows from the special case $\sigma_P = \delta_{P,r} \sigma_P \delta_{P,r}^{-1}$ of the theorem above, then $\delta_{P,s} \delta_{P,r} = \delta_{P,rs}$ holds for all nonzero r and s as well as for the obvious case when both r and s are positive. Thus, $\delta_{P,r}^{-1} = \delta_{P,1/r}$ for any point P and nonzero r. If α is any similarity, then $\alpha \delta_{P,r} \alpha^{-1}$ is a dilatation (Theorem 13.11) fixing point $\alpha(P)$ and has ratio $|r|$. Hence, $\alpha \delta_{P,r} \alpha^{-1} = \delta_{\alpha(P),s}$ where $s = \pm r$. The question is, "Is r the dilation ratio of $\alpha \delta_{P,r} \alpha^{-1}$?" With $P' = \alpha(P)$ and $Q' = \alpha(Q)$ for $Q \neq P$, that the answer is "Yes" follows from the equivalence of each of the following: (1) $r > 0$. (2) $\delta_{P,r}$ is a stretch. (3) $\delta_{P,r}(Q)$ is on $\overrightarrow{PQ}$. (4) $\alpha \delta_{P,r}(Q)$ is on $\overrightarrow{P'Q'}$. (5) $\alpha \delta_{P,r} \alpha^{-1}(\alpha(Q))$ is on $\overrightarrow{P'Q'}$. (6) $\delta_{P,s}(Q')$ is on $\overrightarrow{P'Q'}$. (7) $\delta_{P,s}$ is a stretch. (8) $s > 0$. Since $s = \pm r$ and both r and s have the same sign, then $r = s$, as desired.

Theorem 13.12. *If P is a point, then $\delta_{P,-r} = \sigma_P \delta_{P,r}$ for any $r \neq 0$, $\delta_{P,1} = \iota$, $\delta_{P,-1} = \sigma_P$, and $\delta_{P,s} \delta_{P,r} = \delta_{P,rs}$ for any nonzero r and s. If $\delta_{p,r}$ is a dilation and α is any similarity then*

$$\alpha \delta_{P,r} \alpha^{-1} = \delta_{\alpha(P),r}.$$

If $r \neq 1$, then the nonidentity dilation $\delta_{P,r}$ is said to have **center P**.

Further results on the dilatations are more easily obtained by using coordinates. In the Cartesian plane with $O = (0, 0)$, we clearly have $\delta_{O,r}((x, y)) = (rx, ry)$ for positive r and this same equation must hold for negative r since $\sigma_O((x, y)) = (-x, -y)$. So $\delta_{O,r}$ has equations $x' = rx$ and $y' = ry$ in any case. Now, suppose $P = (a, b)$ and $\delta_{P,r}((x, y)) = (x', y')$. Then, from the equations

$$\delta_{P,r} = \tau_{O,P} \delta_{O,r} \tau_{O,P}^{-1} = \tau_{O,P} \delta_{O,r} \tau_{P,O},$$

we have

$$\delta_{P,r}((x, y)) = (r(x - a) + a, r(y - b) + b) = (x', y')$$

and the following.

Theorem 13.13. *If $P = (a, b)$, then $\delta_{P,r}$ has equations*

$$\begin{cases} x' = rx + (1 - r)a, \\ y' = ry + (1 - r)b. \end{cases}$$

The next theorem follows from the theorem above, and the proofs are left for Exercise 13.2.

Theorem 13.14. *Given* $\delta_{A,\,1/r}$ *and* $\delta_{B,\,r}$, *then for some point* C

$$\delta_{B,\,r}\delta_{A,\,1/r} = \tau_{A,\,C}.$$

Given $\delta_{A,\,r}$ *and* $\delta_{B,\,s}$ *with* $rs \neq 1$, *then for some point* C

$$\delta_{B,\,s}\delta_{A,\,r} = \delta_{C,\,rs}.$$

Given $\tau_{A,\,B}$ *and* $\delta_{A,\,r}$ *with* $r \neq 1$, *then for some point* C

$$\tau_{A,\,B}\delta_{A,\,r} = \delta_{B,\,r}\tau_{A,\,B} = \delta_{C,\,r}.$$

Although the coordinate proofs for Theorem 13.14 are easy to give and the content of the equations themselves is easy to understand, in one sense visualization is virtually impossible. With the help of Figure 13.6 it may be easy to see point for point that the result of $\delta_{A,\,4}$ followed by $\delta_{B,\,1/4}$ is $\tau_{P,\,Q}$. However, it would take a very special mind to see this transformation as the result of the continuous change associated with the dilation about A of ratio 4 followed by that associated with the dilation about B of ratio 1/4. Nevertheless, we know as a result of our proofs that such is the case.

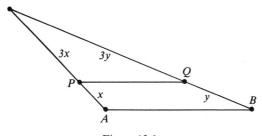

Figure 13.6

A similarity on the Cartesian plane is a stretch about the origin O followed by an isometry (Theorem 13.7). From this fact and the equations for an isometry given by Theorem 9.3, it follows that a similarity has equations of the form

$$\begin{cases} x' = \quad (r\cos\Theta)x - (r\sin\Theta)y + c, \\ y' = \pm[(r\sin\Theta)x + (r\cos\Theta)y + d], \end{cases}$$

where r and Θ are numbers with $r > 0$ and, conversely, equations of this form are those of a similarity. With $a = r\cos\Theta$ and $b = r\sin\Theta$, we have the following.

Theorem 13.15. *A similarity on the Cartesian plane has equations of the form*

$$\begin{cases} x' = \quad ax - by + c, \\ y' = \pm\,[bx + ay + d], \end{cases} \quad \text{with } a^2 + b^2 \neq 0,$$

and, conversely, equations of this form are those of a similarity.

A similarity α that is a stretch about some point P followed by an even isometry is said to be **direct**; a similarity α that is a stretch about some point P followed by an odd isometry is said to be **opposite**. From the equations for isometries and similarities it is evident that whether a similarity is direct or opposite is independent of the point P above. In the equations in Theorem 13.15, the positive sign applies to direct similarities and the negative sign applies to opposite similarities.

Theorem 13.16. *Every similarity is either direct or opposite but not both. The direct similarities form a group. The product of two opposite similarities is direct. The product of a direct similarity and an opposite similarity is an opposite similarity.*

The next chapter contains some applications of similarity theory to plane geometry.

§13.3 Exercises

13.1. Finish the proof of the Classification Theorem for Similarities on the Plane.

13.2. Prove Theorem 13.14.

13.3. Show that $\alpha \rho_{P,\Theta} \alpha^{-1} = \rho_{\alpha(P),\,\pm\Theta}$ for any rotation $\rho_{P,\Theta}$ where the positive sign applies when α is a direct similarity and the minus sign applies when α is an opposite similarity.

13.4. List all similarities whose square is a dilatation.

13.5. Find all fixed points and all fixed lines of $\sigma_l \delta_{A,2}$ where $\triangle ABC$ is equilateral and $l = \overleftrightarrow{BC}$.

13.6. For what point P does a dilation about P have equations $x' = -2x + 3$ and $y' = -2y - 4$?

13.7. Show that any two parabolas are similar.

13.8. What are the fixed points and fixed lines of a stretch reflection? What are the fixed points and fixed lines of a stretch rotation?

13.9. True or False
    (a) A similarity that is not an isometry has a fixed point, and a dilatation that is not a translation has a fixed point.
    (b) $\delta_{C,r}$ fixes point C and otherwise sends point P to P' where P' is the unique point on $\overleftrightarrow{CP}$ such that $\underline{CP'} = r\underline{CP}$.
    (c) The group of all dilatations is generated by the dilations.
    (d) $\sigma_P \delta_{P,r} = \delta_{P,r} \sigma_P$ for any point P and nonzero number r.
    (e) $\delta_{A,r}(B)$ is on $\overrightarrow{AB}$ if $A \neq B$.
    (f) If α is an opposite similarity of ratio r, P is any point, and l is any line, then there is an even isometry β such that $\alpha = \beta \sigma_l \delta_{P,r}$.

(g) $\alpha \tau_{A,B} \alpha^{-1} = \tau_{\alpha(A), \alpha(B)}$ for any similarity α and points A and B.

(h) There is a unique point Q on $\overleftrightarrow{AB}$ such that $AQ/QB = 7$.

(i) A dilatation is a similarity.

(j) A stretch rotation is not an isometry, and neither a stretch nor a rotation is a stretch rotation.

13.10. To determine the height of an object you can place a mirror on the ground and move back until you see the top of the object in the mirror. Explain how this procedure works.

13.11. Given a figure consisting of distinct points A and B, sketch the set of all lines fixed by $\delta_{B,3} \delta_{A,-2}$.

13.12. Prove or disprove: If α is a transformation and δ is a dilation, then $\alpha \delta \alpha^{-1}$ is a dilatation.

13.13. Prove or disprove: If $r > 0$, then a mapping α such that $P'Q' = rPQ$ for all points P and Q with $P' = \alpha(P)$ and $Q' = \alpha(Q)$ is a similarity.

13.14. Suppose α is a transformation such that $\overline{AB} \cong \overline{CD}$ implies $\overline{A'B'} \cong \overline{C'D'}$ for all points A, B, C, D and their images A', B', C', D', respectively, under α. Show α is a similarity.

13.15. Complete each of the following:

(a) If $\delta_{P,3}((x, y)) = (3x + 7, 3y - 5)$, then $P =$ _____ .

(b) If $x' = 3x + 5y + 2$ and $y' = tx - 3y$ are the equations of a similarity, then $t =$ _____ .

(c) If $\sigma_p \delta_{P,15} = \delta_{P,x}$, then $x =$ _____ .

(d) If $\delta_{C,c} \tau_{A,B} = \tau_{P,Q} \delta_{C,c}$, then $P =$ _____ and $Q =$ _____ .

(e) If $\delta_{B,-5} \rho_{A,10} \delta_{B,-5}^{-1} = \rho_{P,10}$, then $P =$ _____ .

(f) If $\delta_{B,s} \delta_{A,t} = \delta_{T,t} \delta_{B,s}$, then $T =$ _____ .

(g) If $\rho_{A,\Theta} \delta_{A,r} = \delta_{A,r} \rho_{A,x}$, then $x =$ _____ .

(h) If $\sigma_p \delta_{P,r} = \delta_{P,r} \sigma_x$ when p is on P, then $x =$ _____ .

(i) If $\tau_{A,B}^{-1} = \tau_{A,C}$, then $C =$ _____ .

13.16. Prove or disprove: A finite group of similarities is either C_n or D_n for some n.

13.17. Prove or disprove: Nonidentity dilatations α and β commute iff α and β are translations.

13.18. Given line l and points C, A, B in Figure 13.7 with $\delta_{C,r}(A) = B$, find M and m such that $\sigma_l \delta_{C,r} = \sigma_m \delta_{M,r}$ and point M is on line m.

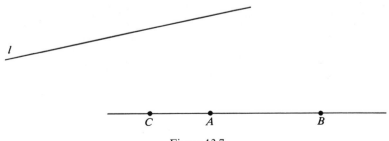

Figure 13.7

13.19. Find all stretch reflections taking point A to point B.

13.20. Find all stretch rotations taking point A to point B.

13.21. If $\alpha((1, 2)) = (0, 0)$ and $\alpha((3, 4)) = (3, 4)$, then what is the ratio of similarity α?

13.22. If $\alpha((0, 0)) = (1, 0)$, $\alpha((1, 0)) = (2, 2)$, and $\alpha((2, 2)) = (-1, 6)$ for similarity α, then find $\alpha((-1, 6))$.

13.23. Prove or disprove: There are exactly two dilatations taking circle A_B to circle C_D.

13.24. Show that if α is a collineation that preserves the set of all circles, then α is a similarity.

13.25. Show that an involutory similarity is a reflection or a halfturn.

13.26. Prove the last equation in Theorem 13.12 by using the fact that $\delta_{Q,r}$ is a square when $r > 0$.

13.27. Show that a nonidentity dilation with center P commutes with σ_l iff P is on l. Show nonidentity dilations $\delta_{A,a}$ and $\delta_{B,b}$ commute iff $A = B$. Show dilatations $\delta_{A,a}$ and $\tau_{A,B}$ never commute if $A \neq B$ and $a \neq 1$.

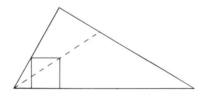

Figure 13.8

13.28. Given acute triangle $\triangle ABC$, construct the square inscribed in the triangle that has a side on $\overline{AB}$. See Figure 13.8 for an idea.

Chapter 14

Classical Theorems

§14.1 Menelaus, Ceva, Desargues, Pappus, Pascal

For each of the mathematicians that lends his name to the title of this section, there is at least one famous theorem that bears his name. As we shall see, the Alexandrian Greek mathematician Pappus can be paired with either of the seventeenth-century French mathematicians Desargues or Pascal. However, the Alexandrian Greek mathematician Menelaus and the seventeenth-century Italian mathematician Ceva are invariably mentioned together. Menelaus' Theorem, which involves a test for the collinearity of three points, and Ceva's Theorem, which involves a test for the concurrency of three lines, are frequently called the *Twin Theorems*. These theorems should have been discovered together, and it is not insignificant that such a long period separates Menelaus and Ceva. During the 1500 years that separate the two there was little development in mathematics.

About the year A.D. 100, Menelaus of Alexandria extended a then well-known lemma to spherical triangles in his *Sphaerica*, the high point of Greek trigonometry. It is this lemma for the plane that today bears the name of Menelaus. Of the Ceva brothers, the lesser known Tommaso (1648–1737) wrote on the cycloid while Giovanni (1647–1736) resurrected the forgotten Menelaus' Theorem and published it in 1678 along with the twin theorem now known as Ceva's Theorem.

Our classical theorems will depend mostly on the following lemma. Read the statement and see if you can prove each of the four parts before you read the proofs given after the statement.

Theorem 14.1. *A dilatation with a fixed point off a fixed line is the identity.* *If $A \neq B$ and $ab \neq 1$, then $\delta_{B,b}\delta_{A,a}$ is a dilation with center on $\overleftrightarrow{AB}$. If*

$$\delta_{R,r}\delta_{Q,q}\delta_{P,p} = \imath$$

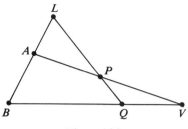

Figure 14.1

and one of the numbers p, q, r is different from 1, then points P, Q, R, are collinear.
If distinct lines $\overleftrightarrow{AB}$ and $\overleftrightarrow{PQ}$ intersect at point L, then

$$\delta_{A,a}(P) = \delta_{B,b}(Q) \text{ implies } \delta_{B,b}^{-1}\delta_{A,a} = \delta_{L,a/b}.$$

A dilatation with a fixed point is a dilation (Theorem 13.5). A dilation fixes its center and all lines through the center, and a dilation fixing any other point or any other line must fix two points and so be the identity (Theorem 13.3). This proves the first part of the theorem. For the second part, if $ab \neq 1$, then the dilatation $\delta_{B,b}\delta_{A,a}$ is a nonidentity dilation that fixes $\overleftrightarrow{AB}$. By the first part, the center of the dilation must be on $\overleftrightarrow{AB}$.

For the third part of Theorem 14.1, suppose $\delta_{R,r}\delta_{Q,q}\delta_{P,p}$ is the identity with points P, Q, R distinct. Then $\delta_{Q,q}\delta_{P,p} = \delta_{R,r}^{-1}$. If $r = 1$, then $p = q = 1$ also, as we would have to have $\delta_{P,p} = \delta_{Q,q}^{-1}$ with $P \neq Q$. So $r \neq 1$ and the center R of dilation $\delta_{Q,q}\delta_{P,q}$ must be on the fixed line $\overleftrightarrow{PQ}$ by the second part of the theorem. To prove the last part of the theorem, let $\delta_{A,a}(P) = \delta_{B,b}(Q) = V$ and $\delta = \delta_{B,b}^{-1}\delta_{A,a}$. See Figure 14.1. Now, $a \neq b$, as otherwise $\underline{AV/AP} = \underline{BV/BQ}$ which holds only if lines $\overleftrightarrow{AB}$ and $\overleftrightarrow{PQ}$ are parallel. So $a/b \neq 1$ and δ is a dilation with center on $\overleftrightarrow{AB}$. Since $\delta(P) = Q$, the dilation δ has center on $\overleftrightarrow{PQ}$. Hence, $\delta = \delta_{L,a/b}$, as desired.

Each of the Twin Theorems involves a certain product of ratios of directed distances. Except in the statement of a theorem, this product will be denoted throughout by $*$ to save space. Suppose points D, E, F are respectively on lines $\overleftrightarrow{BC}$, $\overleftrightarrow{AC}$, $\overleftrightarrow{AB}$ and each point is distinct from the vertices of $\triangle ABC$. Then $*$ is defined by the equation

$$* = (\underline{AF/FB})(\underline{BD/DC})(\underline{CE/EA}).$$

The equation for $*$ is easy to remember. See Figure 14.2. Points D, E, F are on the extended sides opposite vertices A, B, C, respectively. Then to form $*$ we start at A and march around the triangle. First, go from A to B by way of F; second, go from B to C by way of D; and, finally, go from C back to A by way of E. Practice forming $*$ for each $\triangle ABC$ illustrated in Figure 14.2. We are ready to state and then prove **Menelaus' Theorem.**

Theorem 14.2. *Suppose points D, E, F are respectively on lines $\overleftrightarrow{BC}$, $\overleftrightarrow{AC}$, $\overleftrightarrow{AB}$ and each is distinct from the vertices of $\triangle ABC$. Then points D, E, F are collinear iff*

$$(\underline{AF/FB})(\underline{BD/DC})(\underline{CE/EA}) = -1.$$

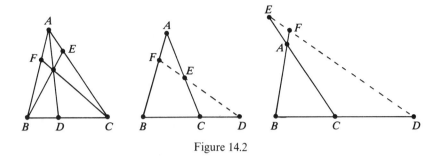

Figure 14.2

For the proof of Menelaus' Theorem, let f, d, e be numbers such that $\delta_{F,f}(B) = A$, $\delta_{D,d}(C) = B$, and $\delta_{E,e}(A) = C$. In other words, numbers f, d, e are such that $\underline{FA} = f\underline{FB}$, $\underline{DB} = d\underline{DC}$, and $\underline{EC} = e\underline{EA}$. Let $\delta = \delta_{F,f}\delta_{D,d}\delta_{E,e}$. Dilatation δ fixes point A and so must be a dilation. Now, first suppose D, E, F are collinear on line l. Then δ has fixed point A off fixed line l. Hence, $\delta = \iota$ and the dilation ratio fde of δ is $+1$. Since $fde = +1$, then $* = -1$. Conversely, suppose $* = -1$. Then $fde = +1$ and $\delta = \iota$. Since none of d, e, f is 1, then points D, E, F are collinear, as desired.

When points D, E, F in Theorem 14.2 are collinear on line t, then t is called a ***transversal*** to $\triangle ABC$. A line through exactly one vertex of a triangle is a ***cevian*** for that triangle. In Figure 14.3, lines $\overleftrightarrow{AD}$, $\overleftrightarrow{BE}$, and $\overleftrightarrow{CF}$ are cevians for $\triangle ABC$. We now state and then prove ***Ceva's Theorem***.

Theorem 14.3. *Suppose points D, E, F are respectively on lines $\overleftrightarrow{BC}$, $\overleftrightarrow{AC}$, $\overleftrightarrow{AB}$ and each is distinct from the vertices of $\triangle ABC$. Then lines $\overleftrightarrow{AD}$, $\overleftrightarrow{BE}$, $\overleftrightarrow{CF}$ are concurrent or parallel iff*

$$(\underline{AF}/\underline{FB})(\underline{BD}/\underline{DC})(\underline{CE}/\underline{EA}) = +1.$$

To begin the proof of Ceva's Theorem, let f, c, d, a be numbers such that

$$\delta_{F,f}(B) = \delta_{C,c}(E) = A \quad \text{and} \quad \delta_{D,d}(B) = \delta_{A,a}(E) = C.$$

So $f = \underline{FA}/\underline{FB}$, $c = \underline{CA}/\underline{CE}$, $d = \underline{DC}/\underline{DB}$, $a = \underline{AC}/\underline{AE}$, and $(fa)/(dc) = *$. Also, note that $f = c$ iff $\overleftrightarrow{FC} \parallel \overleftrightarrow{BE}$ and that $d = a$ iff $\overleftrightarrow{DA} \parallel \overleftrightarrow{BE}$. Let $\delta_1 = \delta_{C,c}^{-1}\delta_{F,f}$ and $\delta_2 = \delta_{A,a}^{-1}\delta_{D,d}$. So $\delta_1(B) = \delta_2(B) = E$. Now, if the cevians

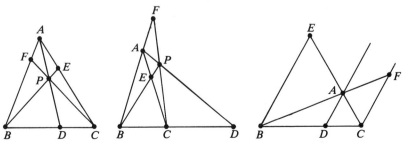

Figure 14.3

$\overleftrightarrow{AD}, \overleftrightarrow{BE}, \overleftrightarrow{CF}$ are parallel, then $f = c, d = a$, and so $* = +1$. If the cevians are concurrent at some point P, then $\delta_1 = \delta_{P, f/c}$ and $\delta_2 = \delta_{P, d/a}$. Since $\delta_1(B) = \delta_2(B) = E$ and $\delta_1(P) = \delta_2(P) = P$, then $\delta_1 = \delta_2$. So $f/c = d/a$ and $* = +1$. Conversely, suppose $* = +1$. Then $f/c = d/a$. If $f/c = d/a = 1$, then $f = c$, $d = a$, and the cevians are parallel. If $f/c = d/a \neq 1$, then δ_1 is a dilation with center Q, the intersection of lines $\overleftrightarrow{CF}$ and $\overleftrightarrow{BE}$, and δ_2 is a dilation with center R, the intersection of lines $\overleftrightarrow{AD}$ and $\overleftrightarrow{BE}$. Thus, $\underline{QE/QB} = \underline{RE/RB}$ because $\delta_1(B) = \delta_2(B) = E$ and so δ_1 and δ_2 have the same dilation ratio. Therefore, since Q and R are on $\overleftrightarrow{BE}$, we have $Q = R$ and the cevians are concurrent (Theorem 13.8). Hence, the cevians are either parallel or concurrent, as desired.

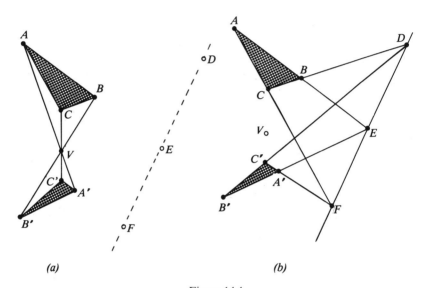

(a)   (b)

Figure 14.4

In Figure 14.4a, triangles $\triangle ABC$ and $\triangle A'B'C'$ are said to be *copolar* as the lines $\overleftrightarrow{AA'}, \overleftrightarrow{BB'}, \overleftrightarrow{CC'}$ joining corresponding vertices of the triangles are concurrent at point V. In Figure 14.4b, triangles $\triangle ABC$ and $\triangle A'B'C'$ are said to be *coaxial* as the points D, E, F of intersection of corresponding extended sides of the triangles are collinear on line l. Note that the figures (a) and (b) can be superimposed; this is the idea behind *Desargues' Theorem*.

Theorem 14.4. *Suppose points $A, B, C, A', B', C', D, E, F, V$ are distinct and triangles $\triangle ABC$ and $\triangle A'B'C'$ are such that lines $\overleftrightarrow{BC}$ and $\overleftrightarrow{B'C'}$ intersect at D, lines $\overleftrightarrow{AC}$ and $\overleftrightarrow{A'C'}$ intersect at E, lines $\overleftrightarrow{AB}$ and $\overleftrightarrow{A'B'}$ intersect at F, and lines $\overleftrightarrow{AA'}$ and $\overleftrightarrow{BB'}$ intersect at V. Then lines $\overleftrightarrow{AA'}, \overleftrightarrow{BB'}, \overleftrightarrow{CC'}$ are concurrent iff points D, E, F are collinear.*

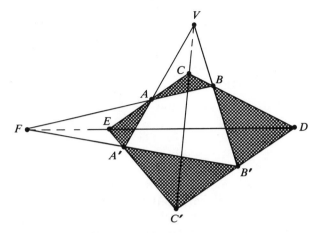

Figure 14.5

First suppose lines $\overleftrightarrow{AA'}$, $\overleftrightarrow{BB'}$, $\overleftrightarrow{CC'}$ in the theorem are concurrent at point V. Let a, b, c be numbers such that

$$\delta_{A,a}(A') = \delta_{B,b}(B') = \delta_{C,c}(C') = V.$$

Then $\delta_{B,b}^{-1}\delta_{A,a} = \delta_{F,a/b}$, $\delta_{C,c}^{-1}\delta_{B,b} = \delta_{D,b/c}$, and $\delta_{A,a}^{-1}\delta_{C,c} = \delta_{E,c/a}$. Hence, $\delta_{E,c/a}\delta_{D,b/c}\delta_{F,a/b} = \iota$ and points D, E, F are collinear, as desired. (This first part alone is sometimes called Desargues' Theorem.)

Conversely, now suppose points D, E, F are collinear. See Figure 14.5. Then $\triangle AEA'$ and $\triangle BDB'$ are copolar from F as $\overleftrightarrow{AB}$, $\overleftrightarrow{ED}$, $\overleftrightarrow{A'B'}$ are concurrent at F. Hence, by the first part above, the triangles are coaxial and points C', V, C must be collinear. In other words, line $\overleftrightarrow{CC'}$ passes through V and, surprisingly, the converse of the first part is actually only a disguised restatement of the first part.

A statement of Desargues' Theorem requires only properties of incidence among the points and lines. This means that the theorem properly belongs to *projective geometry* and is most concisely stated in that context. In projective geometry two lines determine a point as well as two points determine a line: the incidence structure of the points and lines of the Euclidean plane is extended so that any two lines intersect, while Euclidean concepts such as distance and betweenness must be forfeited. We shall not delve into projective geometry except to indicate that some of the mysterious concepts of what goes on "at infinity" can be made precise and meaningful in this geometry. One such idea is that the point V in Theorem 14.4 might be "at infinity." This means that the point V does not exist because lines $\overleftrightarrow{AA'}$, $\overleftrightarrow{BB'}$, $\overleftrightarrow{CC'}$ are parallel. In this case, the theorem takes the following form.

Theorem 14.5. *Suppose $\overleftrightarrow{AA'}$, $\overleftrightarrow{BB'}$, $\overleftrightarrow{CC'}$ are three parallel lines, lines $\overleftrightarrow{BC}$ and $\overleftrightarrow{B'C'}$ intersect at D, lines $\overleftrightarrow{AC}$ and $\overleftrightarrow{A'C'}$ intersect at E, and lines $\overleftrightarrow{AB}$ and $\overleftrightarrow{A'B'}$ intersect at F. Then points D, E, F are collinear.*

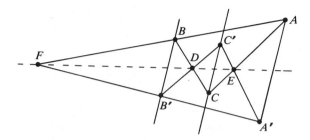

Figure 14.6

The proof of the theorem above uses three dilatations. Since A, B, F are collinear, suppose $\delta_{F,f}(A) = B$. Then $\delta_{F,f}(A') = B'$ since dilation $\delta_{F,f}$ is a dilatation. Since B, C, D are collinear, suppose $\delta_{D,d}(B) = C$. Then $\delta_{D,d}(B') = C'$. Since A, C, E are collinear, suppose $\delta_{E,e}(C) = A$. Then $\delta_{E,e}(C') = A'$. Since the product $\delta_{E,e}\delta_{D,d}\delta_{F,f}$ fixes both A and A', then the product is the identity and D, E, F are collinear, as desired.

In projective geometry, where there are no distinct parallel lines, Desargues' Theorem states that two triangles are co-axial iff they are copolar. Euclidean formulations need to consider the possibility of various lines not intersecting. The following Euclidean formulation states that if two of D, E, F are "at infinity," then so is the third. The proof is left as Exercise 14.1. See Figure 14.7.

Theorem 14.6. *If the three lines $\overleftrightarrow{AA'}$, $\overleftrightarrow{BB'}$, $\overleftrightarrow{CC'}$ are either concurrent or parallel with $\overleftrightarrow{AB} \parallel \overleftrightarrow{A'B'}$ and $\overleftrightarrow{AC} \parallel \overleftrightarrow{A'C'}$, then $\overleftrightarrow{BC} \parallel \overleftrightarrow{B'C'}$.*

It was while Descartes and Fermat were proposing the new methods of analytic geometry that the French engineer and architect Girad Desargues (1591–1661) offered the new field of projective geometry. Desargues' treatise on conics, the *Brouillon Projet* of 1639 that introduced projective geometry, was one of the most unsuccessful great books ever written. In addition to its being revolutionary and extremely terse, the book contained an abundance of new terms. Many of these had a curious botanical nature. For example, a "line" in the new geometry was called a "palm." Of the seventy new terms, one survived mainly because the term was singled out for the sharpest

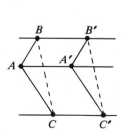

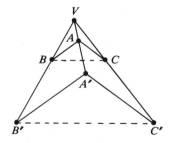

Figure 14.7

ridicule by critics. That term is *involution*. The basic theorem that bears Desargues' name was not published until 1648, where it appears as an appendix by Desargues to a defense by his friend Abraham Bosse of the new type of geometry. For defending Desargues, Bosse lost his job. Desargues' work had been labeled "dangerous and unsound" by his enemies. Desargues was even compelled to post signs all over Paris offering a reward to anyone who could prove his methods were not sound. However, Desargues' ideas were to lie almost dormant until revived in 1822 by Poncelet in his treatise on projective geometry that was written while a prisoner of the Russians during Napoleon's retreat from Moscow.

The theorem named after Pappus can be paired with Desargues' Theorem since the theorems are often encountered together in an axiomatic study of the foundations of geometry. In the history of geometry and mathematics, Descartes and Desargues (circa 1650) are essentially preceded only by Pappus of Alexandria (circa A.D. 320). Although the Museum-Library at Alexandria that began in the third century B.C. with Euclid, Archimedes, and Apollonius continued until the fall of Alexandria in A.D. 641, there was little mathematical creativity in the Western world for 1000 years after the last giant of Greek mathematics, Pappus. From Pappus' great work called *Collection*, it seems that the theorem below that bears the name of Pappus and the theorem above that bears the name of Menelaus were in all probability known to Euclid (circa 300 B.C.).

Pappus' Theorem is another theorem that properly belongs to projective geometry. In that context the theorem states that the three pairs of opposite "sides" of a "hexagon" with alternate vertices on two lines intersect on a line. See Figure 14.8, where the hexagon has vertices A, B, C, D, E, F and the opposite sides intersect in the points L, M, N. Our Euclidean formulation below requires the existence of three additional points of intersection P, Q, R. Thus we next state and then prove a somewhat special case of ***Pappus' Theorem***.

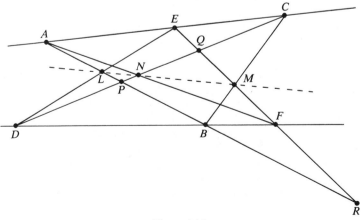

Figure 14.8

Theorem 14.7. *Suppose A, B, C, D, E, F are six points such that points A, C, E are collinear, points B, D, F are collinear, lines $\overleftrightarrow{AB}$ and $\overleftrightarrow{DE}$ intersect at point L, lines $\overleftrightarrow{BC}$ and $\overleftrightarrow{EF}$ intersect at point M, lines $\overleftrightarrow{CD}$ and $\overleftrightarrow{FA}$ intersect at point N, lines $\overleftrightarrow{AB}$ and $\overleftrightarrow{CD}$ intersect at point P, lines $\overleftrightarrow{CD}$ and $\overleftrightarrow{EF}$ intersect at point Q, and lines $\overleftrightarrow{AB}$ and $\overleftrightarrow{EF}$ intersect at point R. Then points L, M, N are collinear.*

Suppose $\delta_{E,e}(R) = \delta_{D,d}(P) = Q$, $\delta_{C,c}(Q) = \delta_{B,b}(R) = P$, and $\delta_{A,a}(P) = \delta_{F,f}(Q) = R$. Then, $\delta_{E,e}^{-1}\delta_{D,d} = \delta_{L,d/e}$, $\delta_{C,c}^{-1}\delta_{B,b} = \delta_{M,b/c}$, and $\delta_{A,a}^{-1}\delta_{F,f} = \delta_{N,f/a}$. Also, $\delta_{L,d/e}(P) = R$, $\delta_{M,b/c}(R) = Q$, $\delta_{N,f/a}(Q) = P$ and so

$$\delta_{N,f/a}\delta_{M,b/c}\delta_{L,d/e}$$

is a dilation with center P and having dilation ratio r with $r = (fbd)/(ace)$. However, $\delta_{C,c}\delta_{E,e}\delta_{A,a} = \imath$ as the product fixes P and $\overleftrightarrow{AC}$, and $\delta_{D,d}\delta_{B,b}\delta_{F,f} = \imath$ as the product fixes Q and $\overleftrightarrow{BD}$. Hence, $cea = 1$, $dbf = 1$, and so $r = 1$. Therefore points L, M, N are collinear, as desired.

We have mentioned that Theorem 14.7 holds if references to P, Q, R are dropped from the statement of the theorem. See Figure 14.9. The same comment applies to Theorem 14.8 below, which is exactly like the statement of Pappus' Theorem except that now the points A, B, C, D, E, F are on a circle. The theorem is sometimes called *Pascal's Theorem* but often goes under the name given by Pascal, the **Mystic Hexagon Theorem**. See Figure 14.10.

Theorem 14.8. *Suppose A, B, C, D, E, F are six points on a circle such that lines $\overleftrightarrow{AB}$ and $\overleftrightarrow{DE}$ intersect at point L, lines $\overleftrightarrow{BC}$ and $\overleftrightarrow{EF}$ intersect at point M, lines*

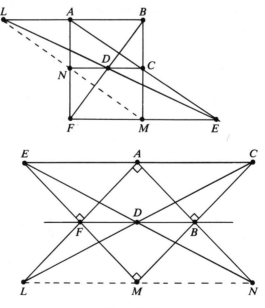

Figure 14.9

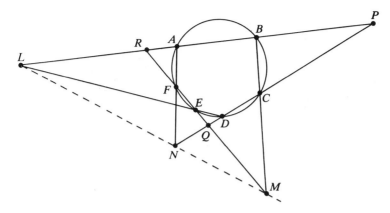

Figure 14.10

$\overleftrightarrow{CD}$ and $\overleftrightarrow{FA}$ intersect at point N, lines $\overleftrightarrow{AB}$ and $\overleftrightarrow{CD}$ intersect at point P, lines $\overleftrightarrow{CD}$ and $\overleftrightarrow{EF}$ intersect at point Q, and lines $\overleftrightarrow{AB}$ and $\overleftrightarrow{EF}$ intersect at point R. Then points L, M, N are collinear.

For the proof of Pascal's Mystic Hexagon Theorem we suppose $a, b, c, d,$ e, f, r are exactly as determined by the beginning of the proof following the statement of Theorem 14.7. (We follow that proof up to the "However.") So $EQ = eER, DQ = dDP, CP = cCQ, BP = bBR, AR = aAP, FR = fFQ,$ and $r = (fbd)/(ace)$. As before, points L, M, N are collinear if $r = 1$. Now,

$$r = \frac{fbd}{ace} = \frac{RE}{RA}\frac{RF}{RB}\ \frac{QC}{QE}\frac{QD}{QF}\ \frac{PA}{PC}\frac{PB}{PD}.$$

Since points A, B, C, D, E, F are on a circle, then each of the three terms in the product on the right has value $+1$. So $r = 1$ and points L, M, N are collinear.

Blaise Pascal (1623–1662) was in on the beginning of the formulation of the theory of probability and is perhaps most often remembered for his extensive study of a certain array that had been known for centuries but that we now call *Pascal's Triangle*. Pascal published his Mystic Hexagon Theorem in 1640, at the age of sixteen. He was a disciple of Desargues and realized that the theorem could be generalized and expanded in projective geometry to state that the three pairs of opposite "sides" of a "hexagon" intersect on one "line" iff the "hexagon" is inscribed in a "conic." This result is often called *Pascal's Theorem* and sometimes the *Pappus–Pascal Theorem* with Pappus' Theorem viewed as the case where the "conic" degenerates to two lines. Pascal is not among the supergiants of mathematics, probably only because he abandoned the subject in favor of theology.

Stated without proof, here are some more properties of the figure associated with Pascal's Mystic Hexagon Theorem. Suppose there are six points on a circle. These points may then be picked in sixty ways to form a "hexagon" inscribed in the circle. Each such hexagon determines a *Pascal line* as given by the theorem. Now, the sixty Pascal lines are concurrent in threes at twenty points, called *Steiner points*, with one Steiner point on each Pascal line. The twenty Steiner points are collinear in fours on fifteen lines, called *Plücker lines*, with each Plücker line through three Steiner points. The sixty Pascal lines are also concurrent in threes at sixty points, called *Kirkman points*, with three Kirkman points on each Pascal line. The sixty Kirkman points are collinear in threes on twenty lines, called *Cayley lines*, with each Cayley line through three Kirkman points and one Steiner point. The twenty Cayley lines are concurrent in fours at fifteen points, called *Salmon points*. Enough?

§14.2 Euler, Brianchon, Poncelet, Feuerbach

This section might well have been called *The Triangle*; we shall look at enough of the classical theorems about the triangle to get an idea of the subject. The nineteenth century saw such a concentrated study of the triangle that the field might seem exhausted. However, even today, new elementary theorems do pop up. Whether developing new results or studying the old results, both professional and amateur have found the triangle a source of delight since the time of Thales.

We are studying $\triangle ABC$. The angle bisectors of the angles at vertex A are necessarily perpendicular. See Figure 14.11. The union of these bisectors is the set of all points equidistant from $\overleftrightarrow{AB}$ and $\overleftrightarrow{AC}$. Suppose that point P is on $\overline{BC}$ and that $\overleftrightarrow{AP}$ is one of these angle bisectors. Then, $\overleftrightarrow{AP}$ is an **angle bisector** of $\triangle ABC$. Also, if $\overleftrightarrow{AQ}$ is perpendicular to $\overleftrightarrow{AP}$, then $\overleftrightarrow{AQ}$ is an **external angle bisector** of $\triangle ABC$. With this notation, as in Figure 14.11, let $\overleftrightarrow{CV}$ be parallel to $\overleftrightarrow{AP}$ with V on $\overleftrightarrow{AB}$. Then, since $AV = AC$, we have $BA/BP = AV/PC = AC/PC$. So $\underline{BP/PC = AB/AC}$. Similarly, for external bisector $\overleftrightarrow{AQ}$ with Q on $\overleftrightarrow{BC}$, we obtain $\underline{BQ/QC = -AB/AC}$. (If $AB = AC$, then the triangle is

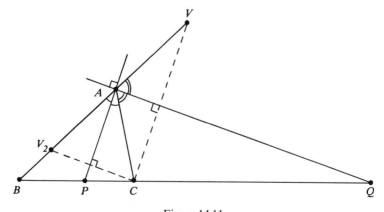

Figure 14.11

isosceles and external angle bisector $\overleftrightarrow{AQ}$ does not intersect $\overleftrightarrow{BC}$.) These same results can be obtained by comparing areas of triangles. In any case, by following Euclid's proof (Proposition VI.3 in Euclid's *Elements*), we have obtained the often overlooked elementary theorem that an angle bisector divides the opposite side of a triangle into segments proportional to the adjacent sides.

Theorem 14.9. *If the angle bisector or the external angle bisector of the angle at A of △ABC intersects $\overleftrightarrow{BC}$ at R, then BR/RC = AB/AC.*

An easy application of Ceva's Theorem and the theorem above gives the following.

Theorem 14.10. *The angle bisectors of a triangle are concurrent. The angle bisector of an angle of a triangle and the two external angle bisectors of the other angles of the triangles are concurrent.*

The point of concurrency of the three angle bisectors of a triangle is the **incenter** of the triangle. In this chapter, I denotes the incenter of △ABC. See Figure 14.12. The point of concurrency of the angle bisector at A and the external angle bisectors at B and C is called an **excenter** of △ABC and is denoted by I_a in this chapter. Excenters I_b and I_c are defined similarly. The incenter and the three excenters are the four points that are equidistant from all three sides of the triangle. So the incenter is the center of the inscribed circle

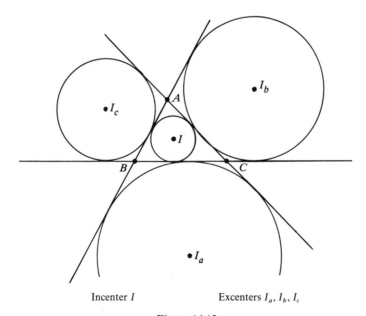

Incenter I Excenters I_a, I_b, I_c

Figure 14.12

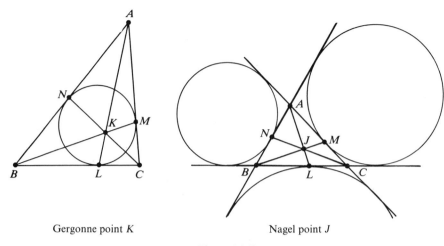

Gergonne point K Nagel point J

Figure 14.13

or *incircle*, which is the circle tangent to the three sides of the triangle. Further, each of the excenters is the center of an escribed circle or *excircle*, which is a circle tangent to a side and to the extensions of the other two sides of the triangle. The incircle and the three excircles for a triangle are illustrated in Figure 14.12.

Since the two tangents to a circle from a point outside the circle are congruent, Ceva's Theorem immediately gives the following theorem. See Figure 14.13.

Theorem 14.11. *The three cevians joining the vertices of a triangle to the points of tangency of the opposite sides with the incircle are concurrent.*

The point of concurrency in Theorem 14.11 is called the ***Gergonne point*** of the triangle. The geometer Joseph-Diaz Gergonne (1771–1859) advocated the methods of analytic geometry and is noted for founding in 1810 the first purely mathematical journal.

Ceva's Theorem and the congruence of tangents to a circle from a point outside the circle also give the next theorem, published in 1836 by C. H. Nagel (1803–1882). The existence of the point of concurrency in the theorem is left for Exercise 14.18. The point of concurrency is called the ***Nagel point*** of the triangle. In Figure 14.13, note that $AB + BL = AC + CL$. So point L is halfway around the perimeter of the triangle from vertex A. Likewise, points M and N are halfway around from vertices B and C, respectively. For this reason, Nagel's Theorem is sometimes called the *Halfway-around Theorem*.

Theorem 14.12. *The cevians joining the vertices of a triangle to the points of tangency of the opposite sides with the corresponding excircles are concurrent.*

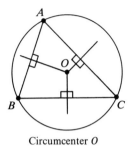
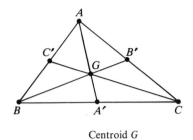

Circumcenter O Centroid G

Figure 14.14

The perpendicular bisector of a side of a triangle is the set of all points equidistant from the endpoints of the side. So the three perpendicular bisectors of the sides of $\triangle ABC$ are concurrent at a point O, which is called the **circumcenter** of the triangle. Since circumcenter O is equidistant from vertices A, B, C, then the circle O_A is the circumscribed circle or **circumcircle** of $\triangle ABC$. See Figure 14.14.

Points A', B', C' will be the midpoints of sides $\overline{BC}$, $\overline{CA}$, $\overline{AB}$, respectively, throughout this section. That the **medians** $\overline{AA'}$, $\overline{BB'}$, $\overline{CC'}$ are concurrent is a trivial result of Ceva's Theorem as $(+1)^3 = +1$. This point of concurrency is the **centroid** G of $\triangle ABC$. See Figure 14.14. Further, by Menelaus' Theorem with $\overleftrightarrow{CC'}$ as a transversal to $\triangle AA'B$, then

$$-1 = (\underline{AG}/\underline{GA'})(\underline{A'C}/\underline{CB})(\underline{BC}/\underline{C'A}) = (\underline{AG}/\underline{GA'})(-1/2)(+1)$$

and so $\underline{AG} = 2\underline{GA'}$. Therefore, the centroid trisects each median. We state these familiar results for the record.

Theorem 14.13. *The perpendicular bisectors of the sides of a triangle are concurrent. The medians of a triangle are concurrent at a point that trisects each median.*

Just for the fun of it, let's get these same results without using the theorems of Menelaus and Ceva. Suppose point G is the intersection of $\overline{BB'}$ and $\overline{CC'}$. Let r be such that $\delta_{G,r}(B) = B'$. Since dilation $\delta_{G,r}$ is a dilatation, then $\delta_{G,r}(C) = C'$, and, since $B'C' = BC/2$, then $r = -1/2$. Since $\delta_{G,r}$ is a dilatation, then $\delta_{G,r}(A)$ is on the line through B' that is parallel to $\overleftrightarrow{AB}$ and also on the line through C' that is parallel to $\overleftrightarrow{AC}$. Hence, $\delta_{G,-1/2}(A) = A'$. Thus, the medians $\overline{AA'}$, $\overline{BB'}$, $\overline{CC'}$ are concurrent at G and $\underline{AG} = 2\underline{GA'}$, as before. Which of the two methods of proof do you like better?

$\triangle A'B'C'$ is called the **medial triangle** of $\triangle ABC$. Points A', B', C' are the respective images of A, B, C under $\delta_{G,-1/2}$. In general, a prime in the remainder of this chapter denotes the image under $\delta_{G,-1/2}$. Since a similarity preserves midpoints, then G' is the centroid of $\triangle A'B'C'$. Hence, since $G = G'$, then the centroid of a triangle is the centroid of its medial triangle. Further, the dilation $\delta_{G,-1/2}$ provides a simple way of showing the altitudes of a

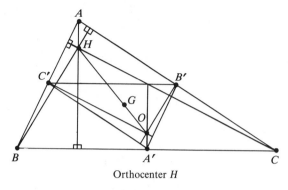

Orthocenter H

Figure 14.15

triangle are concurrent. The ***altitude*** at A of $\triangle ABC$ is the perpendicular to $\overleftrightarrow{BC}$ that passes through A. See Figure 14.15. Under $\delta_{G, -1/2}$, the altitudes of $\triangle ABC$ go to the altitudes of $\triangle A'B'C'$. More interesting is the fact that the image of the altitude at A is also the perpendicular bisector of side $\overline{BC}$. Now we know the perpendicular bisectors of $\triangle ABC$ are concurrent at O. Hence, the altitudes of medial triangle $\triangle A'B'C'$ are concurrent at O. Let H be defined as the point such that $\delta_{G, -1/2}(H) = O$. Then H must be a point on each of the altitudes of $\triangle ABC$. In other words, the altitudes of $\triangle ABC$ are concurrent at point H, called the ***orthocenter*** of $\triangle ABC$.

Theorem 14.14. *The altitudes of a triangle are concurrent. The circumcenter of a triangle is the orthocenter of its medial triangle. The centroid of a triangle is the centroid of its medial triangle.*

Did you notice that the statement of the theorem above did not include a rather surprising result that has already been proved? Since $\delta_{G, -1/2}(H) = O$, then the orthocenter of a triangle, the centroid of the triangle, and the circumcenter of the triangle are on one line. Not only are H, G, O collinear, but G trisects $\overline{HO}$ when $H \neq O$. If $H = O$, then $G = H = O$ and the triangle is equilateral; conversely, if the triangle is equilateral, then points G, H, O coincide. In any case, we have $\underline{HG} = 2\underline{GO}$ since $O = H'$. Under $\delta_{G, -1/2}$, the circumcircle of $\triangle ABC$ and its center O go respectively to the circumcircle of $\triangle A'B'C'$ and its center O'. Letting $N = O'$, we have $\underline{OG} = 2\underline{GN}$. So H, N, G, O are collinear with

$$\underline{HG} = 2\underline{GO} = 4\underline{NG}.$$

In anticipation of a later theorem, we call the circumcircle of the medial triangle of a given triangle the ***ninepoint circle*** of the given triangle. So N is the center of the ninepoint circle of $\triangle ABC$. At present, the midpoints A', B', C' of the sides of $\triangle ABC$ are the only noteworthy points we know to lie on the ninepoint circle of $\triangle ABC$. There will be more. In any case, we know that N is

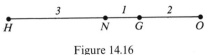

Figure 14.16

the midpoint of the orthocenter H and circumcenter O of $\triangle ABC$. See Figure 14.16.

Theorem 14.15. *The orthocenter H, the center N of the ninepoint circle, the centroid G, and the circumcenter O of a triangle are collinear with $\underline{HG} = 2\underline{GO} = 4\underline{NG}$.*

If $\triangle ABC$ is not equilateral, then the line containing H, N, G, O is called the **Euler line** of $\triangle ABC$. Since $N = O'$, $G = G'$, and $O = H'$, we have the following corollary immediately.

Theorem 14.16. *The Euler line of a nonequilateral triangle is the Euler line of its medial triangle.*

Catherine I of Russia called the Swiss mathematician Leonhard Euler (1707–1783) to the St. Petersburg Academy, which she established at the instigation of her late husband Peter the Great. Although Catherine I died the day Euler arrived in St. Petersburg (Leningrad), Euler remained at the academy until 1741, when Frederick the Great invited him to the Berlin Academy. Frederick the Great called Euler his "mathematical cyclops," as Euler had lost an eye in 1735. After twenty-five years in Berlin, Euler was eager to respond to the invitation of Catherine the Great to return to St. Petersburg. Euler remained in Russia and had been totally blind for seventeen years before he died at the age of seventy-six.

Euler influenced much of the mathematical notation we use today. He is responsible for the notation for e and i as well as standardizing the notation for π. He also put these together to relate the five most important constants in mathematics:

$$e^{i\pi} + 1 = 0.$$

Euler was the most prolific mathematician of all time. His collected works are expected to run close to seventy-five substantial volumes. This supergiant among mathematicians touched most parts of modern mathematics.

The midpoint between the orthocenter and a vertex of a triangle is called an **Euler point** of the given triangle. The triangle with the three Euler points as vertices is the **Euler triangle** of the given triangle. The triangle whose vertices are the feet of the altitudes of $\triangle ABC$ is the **orthic triangle** of $\triangle ABC$. In 1765, Euler showed that the cirumcircle of the orthic triangle and the circumcircle of the medial triangle coincide. (So the ninepoint circle has at least six noteworthy points on it.) In 1822, a joint paper by the French geometers

Charles-Julian Brianchon (1785–1864) and Jean-Victor Poncelet (1788–1867) showed that the circumcircle of the Euler triangle coincides with the other two circumcircles. Therefore, as you would expect and as we shall prove next, the nine-point circle does pass through nine noteworthy points. Poncelet was introduced briefly in the last section. He gave the main impetus to the revival of projective geometry, having laid the foundations of modern projective geometry while a prisoner of the Russians in the Napoleonic Wars. Brianchon is most famous for the theorem of projective geometry he proved as a student. The theorem is known as *Brianchon's Theorem: The three diagonals of a hexagon circumscribed about a conic are concurrent.*

We have defined the ninepoint circle of $\triangle ABC$ to be the circumcircle of the medial triangle $\triangle A'B'C'$. We first wish to show this circle, which has center N, also passes through the feet of the altitudes. Let F_a be the foot of the altitude through A. So F_a and A' are both on $\overleftrightarrow{BC}$. We wish to show $NF_a = NA'$. We suppose $F_a \neq A'$. See Figure 14.17. Now, points H and F_a are on the perpendicular to $\overleftrightarrow{BC}$ at F_a, and points O and A' are on the perpendicular to $\overleftrightarrow{BC}$ at A'. Then, since N is the midpoint of H and O, point N must be on the perpendicular bisector of segment $\overline{F_a A'}$. So $NF_a = NA'$, as desired. Likewise, $NF_b = NB'$ and $NF_c = NC'$, where F_b and F_c are the feet of the altitudes from B and C, respectively. Therefore, the vertices of the orthic triangle are on the ninepoint circle.

Now let E_a, E_b, E_c be the midpoints between H and A, B, C, respectively. See Figure 14.17. We wish to show these Euler points are also on the ninepoint circle of $\triangle ABC$. Since the product $\delta_{G, -1/2}\delta_{H, 2}$ has dilation ratio -1 and fixes point N, then the product is σ_N. So $\sigma_N(E_a) = A'$, $\sigma_N(E_b) = B'$, and $\sigma_N(E_c) = C'$. Hence, $NE_a = NA'$, $NE_b = NB'$, and $NE_c = NC'$, as desired. We have proved the **Ninepoint Circle Theorem** of Brianchon and Poncelet.

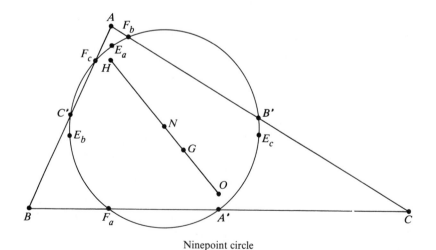

Ninepoint circle

Figure 14.17

Theorem 14.17. *For any triangle the midpoints of the sides, the feet of the altitudes, and the Euler points lie on one circle.*

The name for the ninepoint circle used by English speaking mathematicians obviously comes from the Ninepoint Circle Theorem of Brianchon and Poncelet. Many French mathematicians call the ninepoint circle *Euler's circle*, and the Germans invariably refer to *Feuerbach's circle*. With a drawn sword, Karl Wilhelm Feuerbach (1800–1834) once threatened to behead any of his students who could not solve the equation he had written on the board; the political harassment for erroneous charges concerning his undergraduate activities had taken its toll. Relieved of his teaching duties, the young high-school teacher withdrew from reality and then soon from life itself. However, in a small book published in 1822, Feuerbach had given the world what is often regarded as the most famous of all the theorems of the triangle discovered since the fall of Alexandria in the year 641. Mathematicians from all nations call the result *Feuerbach's Theorem: The ninepoint circle of a triangle is tangent internally to the incircle and tangent externally to each of the excircles of the given triangle.*

Feuerbach's Theorem is stated here without proof. The elementary proofs are too long, and the short proofs involve *inversive geometry*, whose introduction would lead us astray. See page 9 of the Dover reprint of *Circles, A Mathematical View* by D. Pedoe.

The points of tangency for the triangle in Feuerbach's Theorem are called **Feuerbach points** of the given triangle. Feuerbach, as did Euler himself, missed the Euler points on the ninepoint circle. In any case, we now have the original nine plus the four Feuerbach points for a total of thirteen note-worthy points on the ninepoint circle of a scalene triangle.

Starting with $\triangle ABC$ and its orthocenter H, it takes but a minute to see that the orthocenters of $\triangle ABC$, $\triangle HBC$, $\triangle AHC$, $\triangle ABH$ are respectively H, A, B, C. See Figure 14.18. After a couple more minutes, we see that each of these four triangles has the same ninepoint circle and, in fact, the same set

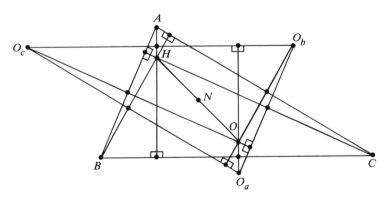

Figure 14.18

of nine distinguished points. Further, let O, O_a, O_b, O_c be the respective images under σ_N of points H, A, B, C. So O is the orthocenter of $\triangle O_a O_b O_c$. Then, by the same argument as above, each of the four triangles $\triangle O_a O_b O_c$, $\triangle O O_b O_c$, $\triangle O_a O O_c$, $\triangle O_a O_b O$ has the same ninepoint circle as the other and with the same set of nine distinguished points. Recall from the proof of Theorem 14.19 that $\sigma_N(E_a) = A'$, $\sigma_N(E_b) = B'$, and $\sigma_N(E_c) = C'$. So the ninepoint circles of $\triangle ABC$ and $\triangle O_a O_b O_c$ coincide. The feet of the altitudes of $\triangle O_a O_b O_c$ provide three additional points to the original nine. Thus all eight triangles have the same ninepoint circle with twelve distinguished points. However, the eight triangles have different incircles and excircles. Hence, each of the eight triangles yields four additional Feuerbach points on the ninepoint circle. Therefore, we have forty-four noteworthy points on the ninepoint circle of $\triangle ABC$. In fact, one can find an infinite sequence of circles tangent to the ninepoint circle to yield an infinite number of "noteworthy" points on the ninepoint circle. Enough!

§14.3 Exercises

14.1. Prove Theorem 14.6.

14.2. Prove part of Ceva's Theorem by using Menelaus' Theorem.

14.3. Prove Desargues' Theorem (Theorem 14.4) by using Menelaus' Theorem. (Note that $\overleftrightarrow{A'B'}$, $\overleftrightarrow{B'C'}$, $\overleftrightarrow{C'A'}$ are respectively transversals to $\triangle ABV$, $\triangle BCV$, $\triangle CAV$.)

14.4. Prove Pappus' Theorem (Theorem 14.7) by applying Menelaus' Theorem three times, with each of $\overleftrightarrow{AF}$, $\overleftrightarrow{BC}$, $\overleftrightarrow{DE}$ as transversals to $\triangle PQR$.

14.5. Suppose $\overleftrightarrow{A_1 A_2}$, $\overleftrightarrow{B_1 B_2 C_1 C_2}$ are three parallel lines such that lines $\overleftrightarrow{B_1 C_1}$ and $\overleftrightarrow{B_2 C_2}$ intersect at point D and lines $\overleftrightarrow{A_1 C_1}$ and $\overleftrightarrow{A_2 C_2}$ intersect at point E. Show that lines $\overleftrightarrow{A_1 B_1}$ and $\overleftrightarrow{A_2 B_2}$ are either parallel to $\overleftrightarrow{DE}$ or else intersect on $\overleftrightarrow{DE}$.

14.6. Prove or disprove: There is a unique dilatation that takes given point P to given point Q and fixes given line $\overleftrightarrow{AB}$.

14.7. Prove or disprove: Given $\triangle ABC$, if point F is on $\overline{AB}$ with $AF = 2FB$, point D is on $\overline{BC}$ with $BD = 2DC$, and point E is on $\overline{AC}$ with $CE = 2AE$, then lines $\overleftrightarrow{AD}$, $\overleftrightarrow{BE}$, $\overleftrightarrow{CF}$ are concurrent.

14.8. Prove Pascal's Mystic Hexagon Theorem (Theorem 14.8) by using Menelaus' Theorem with each of $\overleftrightarrow{BC}$, $\overleftrightarrow{DE}$, $\overleftrightarrow{FA}$ as a transversal to $\triangle PQR$.

14.9. Give a proof of Menelaus' Theorem that is independent of transformation geometry.

14.10. Prove that Ceva's Theorem follows from Menelaus' Theorem.

14.11. Suppose A, B, C, D, E, F are six points such that points A, C, E are collinear, points B, D, F are collinear, lines $\overleftrightarrow{AF}$ and $\overleftrightarrow{CD}$ are parallel, and lines $\overleftrightarrow{BC}$ and $\overleftrightarrow{EF}$ are parallel. Show lines $\overleftrightarrow{AB}$ and $\overleftrightarrow{DE}$ are parallel.

14.12. Show that if cevians $\overleftrightarrow{AD}$, $\overleftrightarrow{BE}$, $\overleftrightarrow{CF}$ of $\triangle ABC$ are concurrent at point Q and D, E, F on $\overleftrightarrow{BC}$, $\overleftrightarrow{CA}$, $\overleftrightarrow{AB}$, respectively, then

$$(AQ/AD) + (BQ/BE) + (CQ/CF) = 2,$$
$$(QD/AD) + (QE/BE) + (QF/CF) = 1.$$

14.13. Suppose $\triangle ABC$ is equilateral. If point E is on side $\overline{AC}$ such that $AE = AC/3$, point D is on side $\overline{BC}$ such that $CD = BC/3$, and $\overline{AD}$ and $\overline{BE}$ intersect at point P, show $\overline{BP}$ and $\overline{CP}$ are perpendicular.

14.14. Given point P in the interior of $\angle CAB$, construct the circles through P that are tangent to both $\overrightarrow{AB}$ and $\overrightarrow{AC}$.

14.15. Given four distinct points A, $\delta_{c,r}(A)$, B, $\delta_{c,r}(B)$ on a line, show how to construct the point C.

14.16. Show how to construct a quadrilateral from four given segments such that the vertices of the quadrilateral will lie on some circle (not given).

14.17. Prove or disprove: For an excircle of a given triangle the lines joining the points of tangency with the extended sides to the opposite vertices are concurrent.

14.18. Prove the existence of the Nagel point of a triangle (Theorem 14.12).

14.19. Show that the external bisectors of the angles of a scalene triangle intersect the opposite sides in three collinear points.

14.20. Show the three lines through the midpoints of the sides of a triangle and parallel to the bisectors of the opposite angles meet in a single point.

14.21. If points P and Q on side $\overline{BC}$ of $\triangle ABC$ have the same midpoint as B and C, then $\overleftrightarrow{AP}$ and $\overleftrightarrow{AQ}$ are called *isotomic conjugate lines* of the triangle. Show that if three cevians, one from each vertex of a triangle, are concurrent at point V, then their isotomic conjugates are concurrent at some point W. The points V and W are called *isotomic conjugate points*.

14.22. Show that the tangents to the circumcircle at the vertices of a scalene triangle intersect the opposite sides in three collinear points.

14.23. Show the feet of the perpendiculars to the sides of a triangle from a point on its circumcircle are collinear. (This is the *Simson Line Theorem*, erroneously named after Robert Simson (1687–1768) but due to William Wallace (1768–1843).)

14.24. Prove or disprove: $\triangle I_a I_b I_c$ has orthocenter I and ninepoint circle that is the circumcircle of $\triangle ABC$.

14.25. Show the incenter of a triangle is the Nagel point of its medial triangle.

14.26. The tangents to the circumcircle at the vertices of a nonright triangle form a triangle called the *tangential triangle* of the given triangle. Show the circumcenter of the tangential triangle of a given triangle lies on the Euler line of the given triangle.

14.27. Given $\triangle ABC$, if R is the circumradius, r is the inradius, r_a is the radius of the excircle with center I_a, s is half the perimeter, and K is the area, then show

$$K = \sqrt{s(s - AB)(s - BC)(s - CA)} = rs$$

$$= \sqrt{rr_ar_br_c} = \frac{(AB)(BC)(CA)}{4R},$$

$$r_a + r_b + r_c = r + 4R.$$

14.28. If lines l and m passing through a vertex of a given triangle are symmetric with respect to the angle bisector at that vertex, then the lines are called *isogonal conjugates* of each other for the triangle. The isogonal conjugate of a median is called a *symmedian* of the triangle. Using the Law of Sines, show a symmedian through a given vertex of a triangle divides the opposite side into segments proportional to the squares of the lengths of the adjacent sides. Then show that the isogonal conjugates of three lines concurrent at a point V off the circumcircle of the triangle, one line through each vertex, are concurrent at a point W. The points V and W are called *isogonal conjugate points*. The isogonal conjugate of the centroid is called the *symmedian point* of the triangle. Show the circumcenter and orthocenter are isogonal conjugate points.

14.29. Show $\underline{JG} = 2\underline{GI}$ for $\triangle ABC$. So $\delta_{G, -1/2}(J) = I$. That is, the incenter of a triangle is the Nagel point of its medial triangle. Define point S by $\delta_{G, -1/2}(I) = S$. See Figure 14.19. Define the *Spieker circle* of $\triangle ABC$ to be the incircle of the medial

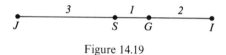

Figure 14.19

triangle. (Recall the Euler circle is the circumcircle of the medial triangle.) Show the Spieker circle has center S and is the incircle of the triangle whose vertices are the three points midway between the Nagel point and the vertices of the triangle.

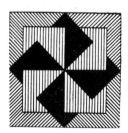

Chapter 15

Affine Transformations

§15.1 Collineations

Up to this point we have studied in modern format mostly the geometry of Euclid. We now turn to transformations that were first introduced by the great mathematician Leonhard Euler (1707–1783). (Euler was introduced in the preceding chapter following Theorem 14.16.) From the meaning of the word *affine*, we must define an ***affine transformation*** as a collineation on the plane that preserves parallelness among lines. So, if l and m are parallel lines and α is an affine transformation, then lines $\alpha(l)$ and $\alpha(m)$ are parallel. However, if β is any collineation and l and m are distinct parallel lines, then $\beta(l)$ and $\beta(m)$ cannot contain a common point $\beta(P)$ as point P would then have to be on both l and m. Therefore, every collineation is an affine transformation. Hence, affine transformations and collineations are exactly the same thing for the Euclidean plane.

Theorem 15.1. *A collineation is an affine transformation; an affine transformation is a collineation.*

The choice between the terms *affine transformation* and *collineation* is sometimes arbitrary and sometimes indicates a choice of emphasis on parallelness of lines or on collinearity of points.

Loosely speaking, *affine geometry* is what remains after surrendering the ability to measure length (isometries) and surrendering the ability to measure angles (similarities) but maintaining the incidence structure of points and lines (collineations). Isometries and similarities are affine transformations since they preserve parallelness. For an example of an affine transformation

167

that is not a similarity, consider the mapping with equations $x' = 2x$ and $y' = y$. In the next section, we shall find a coordinate characterization for the affine transformations. Since we have been studying collineation geometry all along and since collineation geometry and affine geometry are the same thing, then like Molier's character who is pleasantly surprised to find he has been speaking *prose* all his life, you may be pleasantly surprised to find you have been learning *affine geometry* most of your life.

The word *symmetry* probably brings to mind such general ideas as balance, agreement, order, and harmony. Color, sound, and time are often components of symmetry in everyday life. We have been exceedingly conservative in our use of the word *symmetry*; for us, symmetries are restricted to isometries. With a broader mathematical usage of the term, we would certainly be saying that the similarities are the symmetries of similarity geometry and that the collineations are the symmetries of affine geometry. It is in this sense that the entire book is about symmetry. In the most broad usage, the group of all transformations on a structure that preserves the essence of that structure constitutes the symmetries of the structure.

A collineation preserves collinearity of points. We wish to show that, conversely, a transformation such that the images of every three collinear points are themselves collinear must be a collineation. So we suppose α is a transformation that preserves collinearity and aim to show $\alpha(l)$ is a line whenever l is a line. Let A and B be distinct points on l, and let m be the line through points $\alpha(A)$ and $\alpha(B)$. By the definition of α, all the points of $\alpha(l)$ are on m. However, are all the points of m on $\alpha(l)$? Suppose C' is a point on m distinct from $\alpha(A)$ and $\alpha(B)$, and let C be the point such that $\alpha(C) = C'$. To show C must be on l, we assume C is off l and then obtain a contradiction. See Figure 15.1. Now, the image of all the points of $\overleftrightarrow{AB}$, $\overleftrightarrow{BC}$, and $\overleftrightarrow{AC}$ are on m since collinearity is preserved under α. However, any point P in the plane is on a line containing two distinct points of $\triangle ABC$. Since the images of these two points lie on m, then the image of P lies on m. Therefore, the image of every point lies on m, contradicting the fact that α is an onto mapping. Hence, C must lie on l, $m = \alpha(l)$, and α is a collineation, as desired.

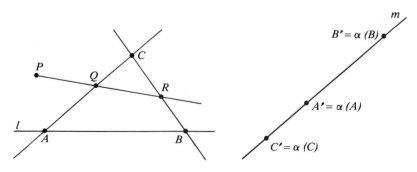

Figure 15.1

Theorem 15.2. *A transformation such that the images of every three collinear points are themselves collinear is an affine transformation.*

Are the affine transformations the same as those transformations for which the images of any three noncollinear points are themselves noncollinear? We shall see the answer is "Yes." Suppose α is an affine transformation. Then α^{-1} is an affine transformation and can't take three collinear points to three noncollinear points. Hence, α can't take three noncollinear points to three collinear points. Therefore, affine transformation α must take any three noncollinear points to three noncollinear points. (Warning, don't be surprised if the following delightful, short proof of the converse of the preceding statement is at first as confusing as suddenly finding yourself at a teddy bears' picnic. The proof of the *converse* follows by the method of proof by *contradiction* while using the *contrapositive* of a known theorem and the trick just used of learning something about a transformation by looking at its *inverse*.) Conversely, suppose β is a transformation such that the images of any three noncollinear points are themselves noncollinear. Assume β is not an affine transformation. Then β^{-1} is not an affine transformation. By the contrapositive of the preceding theorem, then there are three collinear points whose images under β^{-1} are not collinear. Hence, since β is the inverse of β^{-1}, then there are three noncollinear points whose images under β are collinear, a contradiction. Therefore, β is an affine transformation, and we have proved the following theorem.

Theorem 15.3. *A transformation is an affine transformation iff the images of any three noncollinear points are themselves noncollinear.*

The theorem above does *not* state that the image of a triangle under an affine transformation is necessarily a triangle but states only that the images of the vertices of a triangle are themselves vertices of a triangle. We do not know the image of a segment is necessarily a segment. More fundamental, we do not know that an affine transformation necessarily preserves betweenness. It will take some effort to prove this. We begin by showing that *midpoint* is actually an affine concept, that is, an affine transformation carries the midpoint of two given points to the midpoint of their images.

Suppose A and B are distinct points and α is an affine transformation. Let P be any point off $\overleftrightarrow{AB}$. See Figure 15.2. Let Q be the intersection of the line through A that is parallel to $\overleftrightarrow{PB}$ and the line through B that is parallel to $\overleftrightarrow{PA}$. So $\square APBQ$ is a parallelogram. Let $A' = \alpha(A)$, $B' = \alpha(B)$, $P' = \alpha(P)$, and $Q' = \alpha(Q)$. Since two parallel lines go to two parallel lines under α, then $\square A'P'B'Q'$ is a parallelogram. (We are not claiming $\alpha(\square APBQ) = \square A'P'B'Q'$ but only that A', P', B', Q' are vertices in order of a parallelogram.) Further, M, the intersection of $\overleftrightarrow{AB}$ and $\overleftrightarrow{PQ}$, must go to M', the intersection of $\overleftrightarrow{A'B'}$ and $\overleftrightarrow{P'Q'}$. However, since the diagonals of a parallelogram bisect each other,

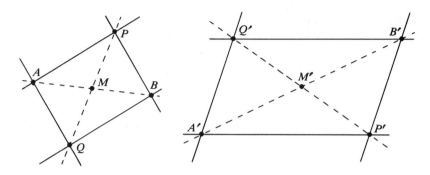

Figure 15.2

then M is the midpoint of A and B while M' is the midpoint of A' and B'. Hence, α preserves midpoints.

Theorem 15.4. *If α is an affine transformation and M is the midpoint of points A and B, then $\alpha(M)$ is the midpoint of $\alpha(A)$ and $\alpha(B)$.*

Suppose α is an affine transformation and the $n + 1$ points P_0, P_1, $P_2, \ldots, P_n$ divide the segment $\overline{P_0 P_n}$ into n congruent segments $\overline{P_{i-1} P_i}$. See Figure 15.3. Let $P_i' = \alpha(P_i)$. Since $P_0 P_1 = P_1 P_2$, $P_1 P_2 = P_2 P_3$, etc., then P_1 is the midpoint of P_0 and P_2, point P_2 is the midpoint of P_1 and P_3, etc. Hence, P_1' is the midpoint of P_0' and P_2', point P_2' is the midpoint of P_1' and P_3', etc. So the images P_0', P_1', $P_2', \ldots, P_n'$ divide the segment $\overline{P_0' P_n'}$ into n congruent segments $\overline{P_{i-1}' P_i'}$.

Theorem 15.5. *If α is an affine transformation, the $n + 1$ points P_0, P_1, $P_2, \ldots, P_n$ divide the segment $\overline{P_0 P_n}$ into n congruent segments $\overline{P_{i-1} P_i}$, and $P_i' = \alpha(P_i)$, then the $n + 1$ points P_0', P_1', $P_2', \ldots, P_n'$ divide the segment $\overline{P_0' P_n'}$ into n congruent segments $\overline{P_{i-1}' P_i'}$.*

It follows from this theorem that P between A and B implies $\alpha(P)$ between $\alpha(A)$ and $\alpha(B)$ provided $\underline{AP/PB}$ is rational. It would have to be a very strange collineation that allowed betweenness not to be preserved in general although preserving midpoints. Early geometers avoided such a monster transformation simply by incorporating the preservation of betweenness within the definition of an affine transformation. In 1880, the French mathematician

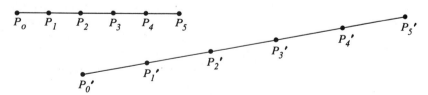

Figure 15.3

Darboux showed the monster transformation does not exist. Gaston Darboux (1842–1917) studied questions of continuity in geometry and analysis. He made contributions to the theory of integration. In particular, he first gave a necessary and sufficient condition for a bounded function to be integrable on a closed interval. (The condition is that the points of discontinuity in the closed interval can be enclosed in a set of intervals whose total length is arbitrarily small.) Darboux taught secondary school from 1867 to 1872 and later held the chair of higher geometry at the Sorbonne. His ingenious proof showing affine transformations preserve betweenness is given here in two parts, the first being a lemma about ratios of directed distances.

Suppose for a positive number t different from 1 that points P and Q are on $\overleftrightarrow{AB}$ and

$$\underline{AP}/\underline{PB} = +t \quad \text{and} \quad \underline{AQ}/\underline{QB} = -t.$$

We intend to show C is the midpoint of P and Q iff

$$\underline{AC}/\underline{CB} = -t^2.$$

First suppose point C is such that this third equation holds. We shall show C is the midpoint of P and Q. The converse will then follow from the uniqueness of a point on $\overleftrightarrow{AB}$ such that the third equation holds. Now,

$$\underline{AP}/\underline{AB} = \underline{AP}/(\underline{AP} + \underline{PB}) = t/(t + 1),$$

$$\underline{AQ}/\underline{AB} = \underline{AQ}/(\underline{AQ} + \underline{QB}) = -t/(-t + 1),$$

$$\underline{AC}/\underline{AB} = \underline{AC}/(\underline{AC} + \underline{CB}) = -t^2/(-t^2 + 1).$$

So

$$(\underline{AP}/\underline{AB}) + (\underline{AQ}/\underline{AB}) = 2(\underline{AC}/\underline{AB}).$$

Hence, $(\underline{AP} + \underline{AQ})/2 = \underline{AC}$ and C must be the midpoint of P and Q.

Theorem 15.6. *If $\underline{AP}/\underline{PB} = +t$ and $\underline{AQ}/\underline{QB} = -t$ for points P and Q on $\overleftrightarrow{AB}$, then C is the midpoint of P and Q iff $\underline{AC}/\underline{CB} = -t^2$.*

For the principal part of Darboux's proof that affine transformations preserve betweenness, suppose α is an affine transformation. Let a prime denote image under α; for example, $P' = \alpha(P)$. Our aim is to show a point between points A and B goes to a point between points A' and B'. If α were to send a point on $\overline{AB}$ to a point off $\overline{A'B'}$, then the affine transformation α^{-1} would send a point off $\overline{A'B'}$ to a point on $\overline{AB}$. Hence, it is sufficient to prove that point C off $\overline{AB}$ implies point C' off $\overline{A'B'}$. (The teddy bears are picnicking again.) We suppose point C is on $\overleftrightarrow{AB}$ but off $\overline{AB}$. Then there is a positive number t different from 1 such that $\underline{AC}/\underline{CB} = -t^2$. Let P and Q be points on $\overleftrightarrow{AB}$ such that $\underline{AP}/\underline{PB} = +t$ and $\underline{AQ}/\underline{QB} = -t$. So C is the midpoint of P and Q by the previous theorem. See Figure 15.4. Now suppose $\overrightarrow{DA} \perp \overleftrightarrow{AB}$. Let E be the point such that A is the midpoint of D and E. Let the line through B

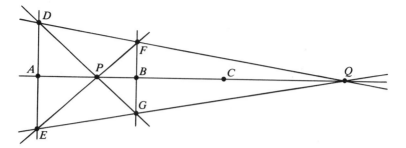

Figure 15.4

that is parallel to $\overleftrightarrow{DE}$ intersect $\overrightarrow{DQ}$ at F and intersect $\overleftrightarrow{EQ}$ at G. Then B is the midpoint of F and G. Since $AD/BF = AQ/BQ = t = AP/PB$, then $\overline{DG}$ and $\overline{EF}$ intersect at P. Consider the image under α of the configuration illustrated in Figure 15.4. We can't expect the perpendicular lines to go to perpendicular lines since angle measure may not be invariant under α. Of course, we can't suppose betweenness is preserved either, since that is what we are proving now. However, since α sends collinear points to collinear points, sends a pair of parallel lines to a pair of parallel lines, and sends midpoints to midpoints, we know quite a bit about the image of the configurative in question. Points A', B', C', P', Q' are collinear. Points D', E', F', G' are off $\overleftrightarrow{A'B'}$ with point A' the midpoint of D' and E', point B' the midpoint of F' and G', and $\overleftrightarrow{D'E'} \parallel \overleftrightarrow{F'G'}$. Three cases might be distinguished according to which of D', F', Q' is between the other two. See Figure 15.5. In any case, by similar triangles there is a nonzero number s (which has absolute value $A'D'/B'G'$ and which is positive iff A'–P'–B') such that $\underline{A'P'/P'B'} = s$ and $\underline{A'Q'/Q'B'} = -s$. Since C' is necessarily the midpoint of P' and Q' then $\underline{A'C'/C'B'} = -s^2$ by the previous theorem. However, since $-s^2$ is negative, then C' is off $\overline{A'B'}$. This finishes Darboux's theorem that affine transformations preserve betweenness.

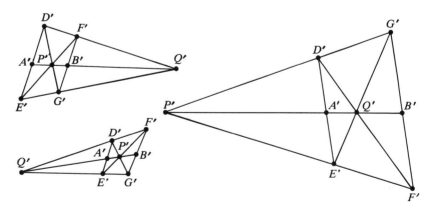

Figure 15.5

Theorem 15.7. *If α is an affine transformation and point P is between points A and B, then point α(P) is between α(A) and α(B).*

Using Theorems 15.5 and 15.7, it is now possible to go back and show $s = t$ in the proof of Theorem 15.7. For this reason, if three points P, A, B are collinear then "AP/PB" is called an *affine ratio* since the value is invariant when the points are replaced by their images under an affine transformation. Thus, the theorems of Menelaus and Ceva from the preceding chapter are often called *affine theorems*. Note that these theorems deal only with the basic incidence structure, which is invariant under collineations.

As an immediate consequence of Theorem 15.7, we know that an affine transformation preserves all those geometric entities whose definition goes back only to the definition of betweenness. Thus, an affine transformation preserves segments, rays, triangles, quadrilaterals, halfplanes, interiors of triangles, etc. The preservation of segments and triangles is singled out for emphasis in the following theorem.

Theorem 15.8. *If A', B', C' are the respective images of three noncollinear points A, B, C, under affine transformation α, then $\alpha(\overline{AB}) = \overline{A'B'}$ and $\alpha(\triangle ABC) = \triangle A'B'C'$.*

Suppose affine transformation α fixes two points A and B. Does α fix $\overleftrightarrow{AB}$ pointwise? Assume there is a point C on $\overleftrightarrow{AB}$ such that $C' \neq C$ with $C' = \alpha(C)$. Without loss of generality, we may suppose C is on $\overrightarrow{AB}$. As an intermediate step, we shall show C is between two fixed points A and D. Let $B_0 = B$ and define B_{i+1} so that B_i is the midpoint of A and B_{i+1} for $i = 0, 1, 2, \ldots$. See Figure 15.6a below. Since A and B_0 are given as fixed by α, then each of $B_1, B_2, B_3, \ldots$ in turn must be fixed by α since α preserves midpoints. Let $D = B_k$ where k is an integer such that $AB_k = 2^k AB > AC$. Then C lies between fixed points A and D. So $\overline{AD}$ is then fixed and both C and C' lie in $\overline{AD}$. Now, let n be an integer large enough so that $nCC' > AD$.

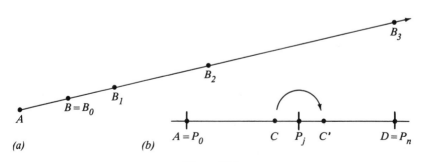

(a)   (b)

Figure 15.6

Let $P_0 = A$, $P_n = D$, and the $n + 1$ points $P_0, P_1, \ldots, P_n$ divide the segment $\overline{AD}$ into n congruent segments $\overline{P_{i-1}P_i}$. See Figure 15.6b. Each of the points P_i is fixed by α by Theorem 15.5. So each $\overline{AP_i}$ and $\overline{P_iD}$ is fixed by α. However, integer n was chosen large enough so that for some integer j point P_j is between C and C'. So C and C' are in different fixed segments $\overline{AP_j}$ and $\overline{P_jD}$, a contradiction. Therefore, $\alpha(C) = C$ for all points on $\overleftrightarrow{AB}$, as desired. The remainder of the following theorem is left for Exercise 15.9.

Theorem 15.9. *An affine transformation fixing two points on a line fixes that line pointwise. An affine transformation fixing three noncollinear points must be the identity. Given $\triangle ABC$ and $\triangle DEF$, there is at most one affine transformation α such that $\alpha(A) = D$, $\alpha(B) = E$, and $\alpha(C) = F$.*

In the next section, we shall see that there is also at least one affine transformation α as described in the last part of the theorem above. Thus an affine transformation will be completely determined once the images of any three noncollinear points are known. The next section also uses the following elementary theorem.

Theorem 15.10. *If (p_1, p_2), (q_1, q_2), and (r_1, r_2) are vertices of a triangle, then the area of that triangle is the absolute value of*

$$[(q_1 - p_1)(r_2 - p_2) - (q_2 - p_2)(r_1 - p_1)]/2.$$

The theorem is proved with reference to the notation in Figure 15.7. You are left to your own devices to prove the area of the shaded triangular region is half the absolute value of $ad - bc$. Since $\tau_{P,O}(P) = (0, 0)$, $\tau_{P,O}(Q) = (a, b)$, and $\tau_{P,O}(R) = (c, d)$, then the area of $\triangle PQR$ is also $|ad - bc|/2$. Substitution then yields the expression given in the statement of the theorem.

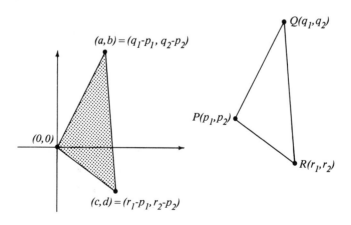

Figure 15.7

§15.2 Linear Transformations

The term "linear transformation" is used in several different contexts in mathematics and so has several different definitions. One thing you can count on is that whenever the term is applied it is applied to something important. For our purposes, a *linear transformation* is any mapping α that has equations

$$\begin{cases} x' = ax + by + h, \\ y' = cx + dy + k, \end{cases} \text{ where } ad - bc \neq 0.$$

The number $ad - bc$ is called the *determinant* of α. A linear transformation is actually a transformation since a given (x, y) obviously determines a unique (x', y') and, conversely, a given (x', y') determines a unique (x, y) precisely because the determinant is nonzero.

As you might expect, linear transformations are related to affine transformations. Let's check that the linear transformation α given above is actually a collineation. Suppose line l has equation $pX + qY + r = 0$. Since p and q are not both zero, then $ap + cq$ and $bp + dq$ are not both zero. So there is a line m with equation

$$(ap + cq)X + (bp + dq)Y + (r + hp + kq) = 0.$$

Line m is introduced because each of the following implies the next where $\alpha((x, y)) = (x', y')$:

(1) (x', y') on line l,
(2) $px' + qy' + r = 0$,
(3) $p(ax + by + h) + q(cx + dy + k) + r = 0$,
(4) $(ap + cq)x + (bp + dq)y + (r + hp + kq) = 0$,
(5) (x, y) on line m.

We have shown α^{-1} is a transformation that takes any line l to some line m. So α^{-1} is a collineation. Hence, α is itself a collineation, proving the first part of the following theorem.

Theorem 15.11. *A linear transformation is an affine transformation; an affine transformation is a linear transformation.*

For the second part of the theorem, suppose α is an affine transformation. Let $\alpha((0, 0)) = (p_1, p_2) = P$, $\alpha((1, 0)) = (q_1, q_2) = Q$, and $\alpha((0, 1)) = (r_1, r_2) = R$. Since $(0, 0), (1, 0), (0, 1)$ are noncollinear, then P, Q, R are noncollinear. Hence the mapping β with equations

$$\begin{cases} x' = (q_1 - p_1)x + (r_1 - p_1)y + p_1, \\ y' = (q_2 - p_2)x + (r_2 - p_2)y + p_2, \end{cases}$$

is a linear transformation, since the absolute value of its determinant is twice the area of $\triangle PQR$ and therefore nonzero (Theorem 15.10). Further, $\beta((0,0)) = \alpha((0,0))$, $\beta((1,0)) = \alpha((1,0))$, and $\beta((0,1)) = \alpha((0,1))$. Therefore (Theorem 15.9), we have $\alpha = \beta$. So α is a linear transformation. This finishes the proof of Theorem 15.11. Choosing the term *linear transformation* over its equivalents *collineation* and *affine transformation* can emphasize a coordinate viewpoint.

Given $\triangle ABC$ and $\triangle DEF$, we know (Theorem 15.9) there is at most one affine transformation α such that $\alpha(A) = D$, $\alpha(B) = E$, and $\alpha(C) = F$. We now show there is at least one such affine transformation α. From the preceding paragraph, we see how to find the equations for a linear transformation β_1 such that $\beta_1((0,0)) = A$, $\beta_1((1,0)) = B$, and $\beta_1((0,1)) = C$. Repeating the process, we see there is a linear transformation β_2 such that $\beta_2((0,0)) = D$, $\beta_2((1,0)) = E$, and $\beta_2((0,1)) = F$. The linear transformation $\beta_2 \beta_1^{-1}$ is the desired affine transformation α that takes points A, B, C to points D, E, F, respectively.

Theorem 15.12. *Given $\triangle ABC$ and $\triangle DEF$, there is a unique affine transformation α such that $\alpha(A) = D$, $\alpha(B) = E$, and $\alpha(C) = F$.*

Some specific, basic linear transformations are introduced next. If for nonzero number k linear transformations α and β have, respectively, sets of equations

$$\begin{cases} x' = x, \\ y' = ky, \end{cases} \quad \text{and} \quad \begin{cases} x' = kx, \\ y' = y, \end{cases}$$

then α is called a **strain** of ratio k about the X-axis and β is called a **strain** of ratio k about the Y-axis. For fixed k, the composite $\beta\alpha$ is the familiar dilation about the origin with dilation ratio k and having equations $x' = kx$ and $y' = ky$. With a **strain** of ratio k about a given line defined analogously (Exercise 15.8), it follows (Exercise 15.19) that any dilation is the product of two strains about perpendicular lines.

The strain with equations $x' = 2x$ and $y' = y$ fixes the Y-axis pointwise and stretches out the plane away from and perpendicular to the Y-axis. See Figure 15.8 where (b) illustrates the images of the elements shown in (a). As with similarity theory, the terminology here is not standardized. Each of the following words has been used for a strain or for a strain with positive ratio: enlargement, expansion, lengthening, stretch, compression. Seek solace in the fact that whatever property is used to define the term *affine transformation*, the term means the same thing the world over.

Another basic affine transformation is a **shear** about the X-axis having equations

$$\begin{cases} x' = x + ky, \\ y' = y. \end{cases}$$

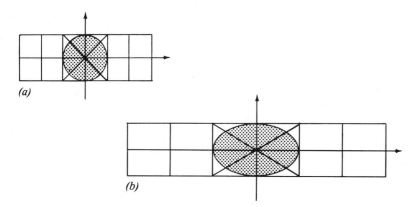

Figure 15.8

Here the X-axis is fixed pointwise and every point is moved horizontally a directed distance proportional to its directed distance from the X-axis. See Figure 15.9 where (b) illustrates the images of the elements shown in (a) for the case $k = 1$. We shall see below that a shear has the property of preserving area. An affine transformation that preserves area is said to be *equiaffine*.

Strains and shears are basic in that simplifying the equations

$$\begin{cases} x' = a\left[\left[x + \dfrac{ab + cd}{a^2 + c^2} y\right]\right] - c\left[\dfrac{ad - bc}{a^2 + c^2} [y]\right] + h, \\ y' = c\left[\left[x + \dfrac{ab + cd}{a^2 + c^2} y\right]\right] + a\left[\dfrac{ad - bc}{a^2 + c^2} [y]\right] + k, \end{cases}$$

we see that the general linear transformation with equations

$$\begin{cases} x' = ax + by + h, \\ y' = cx + dy + k, \end{cases} \quad \text{with } ad - bc \neq 0$$

can be factored into the similarity with equations

$$\begin{cases} x' = ax - cy + h, \\ y' = cx + ay + k, \end{cases}$$

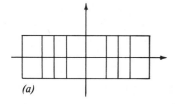

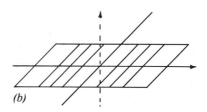

Figure 15.9

following the strain with equations

$$\begin{cases} x' = x, \\ y' = \dfrac{ad - bc}{a^2 + d^2}\, y, \end{cases}$$

following the shear with equations

$$\begin{cases} x' = x + \dfrac{ab + cd}{a^2 + c^2}\, y, \\ y' = \qquad\qquad y. \end{cases}$$

Theorem 15.13. *An affine transformation is the product of a shear, a strain, and a similarity.*

Factoring is hardly ever easy in mathematics. Here you are asked only to check the factoring given next is correct. First, the shear with equations

$$\begin{cases} x' = x + y, \\ y' = \qquad y, \end{cases}$$

can be factored into the composite of the similarity with equations

$$\begin{cases} x' = [(5 - \sqrt{5})/20]\quad x + [(5 - 3\sqrt{5})/20]y, \\ y' = [(-5 + 3\sqrt{5})/20]x + [(5 - \sqrt{5})/20]\ y, \end{cases}$$

following the strain with equations

$$\begin{cases} x' = [(3 + \sqrt{5})/2]x, \\ y' = \qquad\qquad y, \end{cases}$$

following the similarity with equations

$$\begin{cases} x' = \qquad\qquad 2x + (1 + \sqrt{5})\ y, \\ y' = -(1 + \sqrt{5})\ x + \qquad\qquad 2y. \end{cases}$$

Secondly, the nonidentity shear with equations $x' = x + ky$ and $y' = y$ can be factored into the strain of ratio k about the Y-axis ($x' = kx$, $y' = y$) following the shear that was just factored above following the strain of ratio $1/k$ about the Y-axis ($x' = x/k$, $y' = y$). Putting these results together with the previous theorem, we see that an affine transformation is a product of strains and similarities. Since a similarity is an isometry following a dilation about the origin (Theorem 13.7) and since a dilation about the origin is a product of two strains, then an affine transformation is a product of strains and isometries. However, isometries are products of reflections, which are special cases of strains. Thus, affine transformations are products of strains. We have proved the part of the theorem below that is not in parentheses. We shall not take the time to prove the part in parentheses; that result is blatantly stated without proof.

Theorem 15.14. *An affine transformation is a product of strains. (An affine transformation is the product of a strain and a similarity.)*

Suppose $x' = ax + by + h$ and $y' = cx + dy + k$ are equations for affine transformation α. So the determinant $ad - bc$ of α is nonzero. What are the necessary and sufficient conditions for α to be equiaffine? In other words, when is area preserved by α? Suppose P, Q, R are noncollinear points with $P = (p_1, p_2)$, $Q = (q_1, q_2)$, $R = (r_1, r_2)$, $P' = \alpha(P) = (p'_1, p'_2)$, $Q' = \alpha(Q) = (q'_1, q'_2)$, and $R' = \alpha(R) = (r'_1, r'_2)$. Recall (Theorem 15.10) that the area PQR of $\triangle PQR$ is given by

$$PQR = \pm[(q_1 - p_1)(r_2 - p_2) - (q_2 - p_2)(r_1 - p_1)]/2$$

and similarly the area $P'Q'R'$ of $\triangle P'Q'R'$ is given by

$$P'Q'R' = \pm[(q'_1 - p'_1)(r'_2 - p'_2) - (q'_2 - p'_2)(r'_1 - p'_1)]/2.$$

Substitution shows that

$$P'Q'R' = \pm(ad - bc)PQR.$$

Thus, under an affine transformation with determinant t, area is multiplied by $\pm t$. This more than answers our question about preserving area. Area is preserved by α when the determinant of α is ± 1.

Continuing with the same notation for affine transformation α, we recall that α is a similarity iff there is a positive number r such that $P'Q' = rPQ$ for all points P and Q. With substitution, this equation becomes

$$\sqrt{(a^2 + c^2)(q_1 - p_1)^2 + (b^2 + d^2)(q_2 - p_2)^2 + 2(ab + cd)(q_1 - p_1)(q_2 - p_2)}$$
$$= r\sqrt{(q_1 - p_1)^2 + (q_2 - p_2)^2}.$$

This equation can hold for all p_1, p_2, q_1, q_2 iff $a^2 + c^2 = b^2 + d^2 = r^2$ and $ab + cd = 0$. Since a similarity of ratio r is an isometry iff $r = 1$, we have our last theorem.

Theorem 15.15. *Suppose affine transformation α has equations*

$$\begin{cases} x' = ax + by + h, \\ y' = cx + dy + k, \end{cases} \quad \text{with } ad - bc \neq 0.$$

Transformation α is equiaffine iff

$$|ad - bc| = 1.$$

Transformation α is a similarity iff

$$a^2 + c^2 = b^2 + d^2 \quad \text{and} \quad ab + cd = 0.$$

Transformation α is an isometry iff

$$a^2 + c^2 = b^2 + d^2 = 1 \quad \text{and} \quad ab + cd = 0.$$

§15.3 Exercises

15.1. If $x' = 2x$ and $y' = y$ are the equations of transformation α, show α is a collineation that is not a similarity.

15.2. Prove the area of a triangle with vertices $(0, 0)$, (a, b), (c, d) is half the absolute value of $ad - bc$.

15.3. Show that any given ellipse is the image of the unit circle under some affine transformation.

15.4. If $P = (-2, -1)$, $Q = (1, 2)$, and $R = (3, -6)$, what is the area of $\triangle PQR$? What are the areas of the images of $\triangle PQR$ under the collineations α_k and β_k from the next exercise?

15.5. For a given nonzero number k, find all fixed points and fixed lines for the affine transformations α_k and β_k with respective equations

$$\begin{cases} x' = kx, \\ y' = y, \end{cases} \quad \text{and} \quad \begin{cases} x' = x + ky, \\ y' = \quad y. \end{cases}$$

15.6. Show that $\{\alpha_k | k \neq 0\}$ and $\{\beta_k\}$ form abelian groups, when α_k and β_k are defined in the preceding exercise.

15.7. True or False
(a) An affine transformation is a collineation; a collineation is a linear transformation; and a linear transformation is an affine transformation.
(b) An affine transformation is determined once the images of three given points are known.
(c) If $\triangle ABC \cong \triangle DEF$, then there is a unique affine transformation α such that $\alpha(A) = D, \alpha(B) = E$, and $\alpha(C) = F$.
(d) If $\triangle ABC \sim \triangle DEF$, then there is a unique affine transformation α such that $\alpha(A) = D$, $\alpha(B) = E$, and $\alpha(C) = F$.
(e) If $\triangle ABC$ and $\triangle DEF$ are any given triangles, then there is a unique affine transformation such that $\alpha(\triangle ABC) = \triangle DEF$.
(f) Strains and shears are equiaffine.
(g) A shear is a product of strains and similarities.
(h) A collineation is a product of strains and similarities.
(i) A collineation is a product of strains and isometries.
(j) A dilation is a product of strains; a strain is a product of dilations.

15.8. Given nonzero number k and line l, give a definition for the *strain* of ratio k about line l.

15.9. Finish the proof of Theorem 15.9.

15.10. If $x' = ax + by + h$ and $y' = cx + dy + k$ are the equations of mapping α and $ad - bc = 0$, then show α is not a collineation since all images are collinear.

15.11. Prove or disprove: If linear transformation α has determinant t, then α^{-1} has determinant t^{-1}.

15.12. Suppose any affine transformation is the product of a strain and a similarity. Then show that an affine transformation is a product of two strains about perpendicular lines and an isometry. (To see that the perpendicular lines cannot be chosen arbitrarily, see the next exercise.)

15.13. Show the shear with equations $x' = x + y$ and $y' = y$ is not the product of strains about the coordinate axes followed by an isometry.

15.14. Show that the shears do not form a group.

15.15. Prove or disprove: The shears generate the group of affine transformations.

15.16. Prove or disprove: An equiaffine similarity is an isometry.

15.17. Give an example of an equiaffine collineation that is neither an isometry nor a shear.

15.18. Prove or disprove: An involutory affine transformation is a reflection or a halfturn.

15.19. Using your definition from Exercise 15.8, show that a dilation is a product of two strains.

15.20. Find equations for the strain of ratio k about the line with equation $Y = mX$.

15.21. Show Theorem 15.15 is compatible with Theorem 9.3 and with Theorem 13.15.

15.22. Use Theorems 15.5 and 15.7 to show the affine ratio $\underline{AP/PB}$ is invariant under an affine transformation where A, P, B are three collinear points.

15.23. For points P and Q on a line with slope m and for linear transformation α, let $P = (p_1, p_2)$, $Q = (q_1, q_2)$, $P' = \alpha(P)$, and $Q' = \alpha(Q)$. Then show $P'Q' = k_m|q_1 - p_1|$ where k_m is a constant depending on only m and the coefficients in the equations for α. Also, use the result to solve the problem in the preceding exercise.

Figure 15.10

15.24. In Figure 15.10, construct P' where A', B', C', P' are the images of A, B, C, P under a linear transformation.

Chapter 16

Transformations on Three-space

§16.1 Isometries on Space

This chapter may be read following Chapter 13.

Turning to Euclidean three-space, we abandon all of the technical vocabulary and notation from the previous chapters. The reason for this drastic action is that we now want to use these words and symbols in the context of space. So the words and symbols will have new meanings. We suppose from now on that all points, lines, and planes are in three-space. Planes are denoted by capital Greek letters. Of course, we do not abandon all the knowledge from our study of plane collineations. By a careful choice of words we are able to refer to some of these results. However, to avoid total confusion, we agree never to use the symbols for the transformations defined below to indicate their old meaning. The definitions needed are given together for easy reference.

A *transformation* on three-space is a one-to-one correspondence from the set of points in space onto itself. A transformation α having the property that $\alpha(l)$ is a line for every line l is a *collineation*. Transformation α is an *isometry* if $P'Q' = PQ$ for all points P and Q where $P' = \alpha(P)$ and $Q' = \alpha(Q)$. The *identity* ι is defined by $\iota(P) = P$ for every point P. Transformation α is an *involution* iff $\alpha^2 = \iota$ but $\alpha \neq \iota$. Isometry α is a *symmetry* for a set of points if α fixes that set of points.

If Δ is a plane, then the *reflection* σ_Δ is the mapping on the points in space such that $\sigma_\Delta(P) = P$ if point P is on Δ and $\sigma_\Delta(P) = Q$ if point P is off Δ and plane Δ is the perpendicular bisection of $\overline{PQ}$. A product of an even number of reflections is *even*; a product of an odd number of reflections is *odd*. If planes Γ and Δ are parallel, then $\sigma_\Delta \sigma_\Gamma$ is a *translation* along the common perpendicular lines to planes Γ and Δ. If two planes Γ and Δ intersect at line l, then

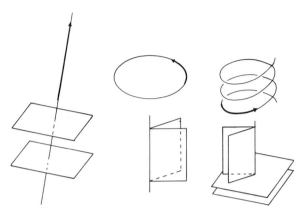

Figure 16.1

$\sigma_\Delta \sigma_\Gamma$ is a **rotation** about axis l. Note that the identity is a translation but not a rotation. The **halfturn** σ_l about line l is the involutory rotation about l. A **screw** is the product of a nonidentity translation along a line l and a rotation about the line l. (A halfturn is sometimes called a *line reflection*, and a screw is sometimes called a *twist* or a *glide rotation*.) See Figure 16.1.

If Γ and Δ are two parallel planes each perpendicular to plane Π, then $\sigma_\Pi \sigma_\Delta \sigma_\Gamma$ is a **glide reflection** with axis Π. If Γ and Δ are two intersecting planes each perpendicular to plane Π, then $\sigma_\Pi \sigma_\Delta \sigma_\Gamma$ is a **rotary reflection** about the point common to Γ, Δ, and Π. If M is a point, the **inversion** σ_M about M is the transformation such that $\sigma_M(P) = Q$ for all points P where M is the midpoint of P and Q. (A rotary reflection is also called a *rotatory reflection*. To distinguish among the many meanings of the word "inversion" in geometry, transformation σ_M is sometimes called a *central inversion* or a *point reflection*.)

Without looking back, define each of the words: transformation, collineation, isometry, and symmetry. Then, look back to check your definitions. Since each plane is the locus of all points equidistant from some two points, then an isometry must preserve the set of planes in space. Since each line is the intersection of distinct intersecting planes, then an isometry must be a collineation. The remainder of the first theorem is immediate.

Theorem 16.1. *An isometry is a collineation; an isometry preserves segments, rays, and planes. The set of all collineations forms a group. The set of all symmetries for a given set of points forms a group. The set of all isometries forms a group.*

Can you define the reflection σ_Δ? Check with the definition above and note that a reflection is an involution. We want to show σ_Δ is also an isometry. If A and B are two points, then there is a plane Γ that contains both A and B and that is perpendicular to plane Δ. See Figure 16.2. (Plane Γ is unique unless $\overleftrightarrow{AB} \perp \Delta$.) Suppose Γ and Δ intersect at line l. By the definition of σ_Δ,

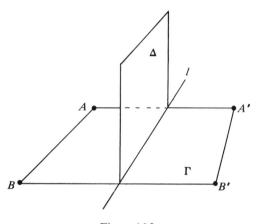

Figure 16.2

points A' and B' are also in Γ where $A' = \sigma_\Delta(A)$ and $B' = \sigma_\Delta(B)$. The restriction of σ_Δ to the plane Γ is just the plane reflection in line l. Hence, $A'B' = AB$ by our results on plane isometries. Therefore, reflection σ_Δ is an isometry.

Theorem 16.2. *A reflection is an involutory isometry. A product of reflections is an isometry.*

The converse of the second statement in the theorem above is our next goal. Analogous to the theorem that a plane isometry is a product of at most three plane reflections, we shall show that an isometry here is a product of at most four reflections. A line having two fixed points under an isometry is fixed pointwise as each point on the line through the two fixed points is uniquely determined by its distances from these two points. A plane having three noncollinear fixed points under an isometry is fixed pointwise as each point on the plane is uniquely determined by its distances from these three noncollinear points (cf. Theorem 5.1). So, if an isometry fixes four points not in one plane, then the isometry fixes pointwise each plane that contains a side of the tetrahedron having the four fixed points as vertices. Hence, since each point in space is on a line that intersects the tetrahedron in two points, then each point is fixed by such an isometry. We have shown an isometry fixing four points not all in one plane must be the identity. Thus, if α and β are isometries that send each of the vertices of some tetrahedron to the same points, then $\beta^{-1}\alpha$ fixes the vertices of the tetrahedron, $\beta^{-1}\alpha = \iota$, and $\alpha = \beta$. We have an analogue of Theorem 5.2.

Theorem 16.3. *If α and β are isometries such that*

$$a(P) = \beta(P), \qquad \alpha(Q) = \beta(Q), \qquad \alpha(R) = \beta(R), \qquad \alpha(S) = \beta(S)$$

for four points P, Q, R, S not all in one plane, then $\alpha = \beta$.

Our goal is still to show that the reflections are building blocks for the isometries. Suppose isometry α takes points P, Q, R, S not all in one plane to points P', Q', R', S', respectively. (Note that α is the unique isometry to do so.) We consider five cases.

Case 1: $P = P'$, $Q = Q'$, $R = R'$, and $S = S'$. We have considered this case above. Here $\alpha = \iota = \sigma_\Delta \sigma_\Delta$ where Δ is any plane through P.

Case 2: $P = P'$, $Q = Q'$, $R = R'$, but $S \neq S'$. Then the plane Γ containing P, Q, R is the perpendicular bisector of $\overline{SS'}$. (Why?) So, $\alpha = \sigma_\Gamma$.

Case 3: $P = P'$, $Q = Q'$, but $R \neq R'$. Here points P and Q lie on the plane Δ that is the perpendicular bisector of $\overline{RR'}$. A reflection in Δ then reduces the problem to Case 1 or to Case 2 above. So α is a product of at most two reflections.

Case 4: $P = P'$ but $Q \neq Q'$. Then point P lies on the plane Π that is the perpendicular bisector of $\overline{QQ'}$ and a reflection in Π reduces the problem to one of the cases above. Hence, α is a product of at most three reflections in planes through P.

Case 5: $P \neq P'$. A reflection in the plane that is the perpendicular bisector of $\overline{PP'}$ reduces the problem to one of the four previous cases. Hence, every isometry is the product of at most four reflections. We have achieved our initial goal.

Theorem 16.4. *An isometry is the product of at most four reflections. An isometry with a fixed point P is the product of at most three reflections in planes through P. An isometry that fixes two points is a rotation, a reflection, or the identity. An isometry that fixes three noncollinear points is a reflection or the identity. An isometry that fixes four nonplanar points is the identity.*

Many theorems for space are evident from our study of plane isometries. Our next theorem lists some of those that follow from our knowledge of plane translations. Be sure you can define a translation. See Figure 16.3.

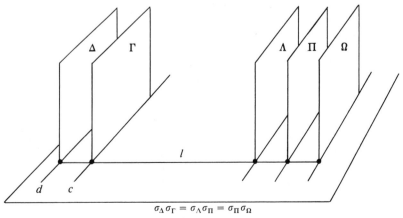

$$\sigma_\Delta \sigma_\Gamma = \sigma_\Lambda \sigma_\Pi = \sigma_\Pi \sigma_\Omega$$

Figure 16.3

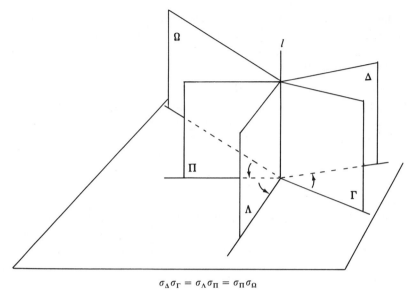

$$\sigma_\Delta \sigma_\Gamma = \sigma_\Lambda \sigma_\Pi = \sigma_\Pi \sigma_\Omega$$

Figure 16.4

Theorem 16.5. *If two planes Γ and Δ are parallel, then the translation $\sigma_\Delta \sigma_\Gamma$ fixes every line that is a common perpendicular to Γ and Δ, fixes every plane that is perpendicular to each of Γ and Δ, but fixes no points. The inverse of a translation is a translation. If the planes Γ, Δ, Π are perpendicular to line l, then there exists unique planes Λ and Ω such that*

$$\sigma_\Delta \sigma_\Gamma = \sigma_\Lambda \sigma_\Pi = \sigma_\Pi \sigma_\Omega.$$

If planes Γ, Δ, Π are perpendicular to line l, then $\sigma_\Pi \sigma_\Delta \sigma_\Gamma$ is a reflection in a plane perpendicular to l. If P and Q are distinct points, then there is a unique translation taking P to Q, and this translation may be expressed as $\sigma_\Delta \sigma_\Gamma$ where either one of Γ or Δ is an arbitrarily chosen plane perpendicular to $\overleftrightarrow{PQ}$ and the other is then a uniquely determined plane perpendicular to $\overleftrightarrow{PQ}$.

The unique translation taking point P to point Q is denoted by $\tau_{P,Q}$. The next theorem lists some of the results that follow from our knowledge of plane rotations. Be sure you can define a rotation. See Figure 16.4.

Theorem 16.6. *If two planes Γ and Δ intersect at line l, then the rotation $\sigma_\Delta \sigma_\Gamma$ fixes exactly those points that are on l, fixes every plane that is perpendicular to l, and fixes every circle having center on l in such a plane. If planes Γ, Δ, Π are concurrent at every point on line l, then there are unique planes Λ and Ω on l such that*

$$\sigma_\Delta \sigma_\Gamma = \sigma_\Lambda \sigma_\Pi = \sigma_\Pi \sigma_\Omega.$$

If planes Γ, Δ, Π are concurrent at each point on line l, then $\sigma_\Pi \sigma_\Delta \sigma_\Gamma$ is a reflection in a plane containing l. A rotation about line l may be expressed as $\sigma_\Delta \sigma_\Gamma$ where either one of Γ or Δ is an arbitrarily chosen plane containing l and the other is then a uniquely determined plane containing l.

If α is any isometry and Γ is any plane, then $\alpha \sigma_\Gamma \alpha^{-1}$ is an involutory iso-metry that fixes every point on plane $\alpha(\Gamma)$ and so must be the reflection in $\alpha(\Gamma)$. This result and our knowledge of plane halfturns lead to the theorem below. Be sure you can define a halfturn.

Theorem 16.7. *If α is any isometry and Γ is any plane, then*

$$\alpha \sigma_\Gamma \alpha^{-1} = \sigma_{\alpha(\Gamma)}.$$

If Γ and Δ are planes, then $\sigma_\Delta \sigma_\Gamma = \sigma_\Gamma \sigma_\Delta$ iff $\Gamma = \Delta$ or $\Gamma \perp \Delta$. If Γ and Δ are any planes perpendicular at line l, then $\sigma_l = \sigma_\Delta \sigma_\Gamma$. If lines p, q, r are perpen-dicular to plane Π, then $\sigma_r \sigma_q \sigma_p$ is a halfturn about a line perpendicular to Π and $\sigma_r \sigma_q \sigma_p = \sigma_p \sigma_q \sigma_r$. If planes Γ, Δ, Π have a common line or a common perpendicular line, then $\sigma_\Pi \sigma_\Delta \sigma_\Gamma = \sigma_\Gamma \sigma_\Delta \sigma_\Pi$.

Be sure you can define an inversion. Let Γ, Δ, Π be three mutually perpen-dicular planes concurrent at point M. For any point P different from M, points P and $\sigma_\Pi \sigma_\Delta \sigma_\Gamma(P)$ are the endpoints of a diagonal of a rectangle with center M. See Figure 16.5 where $P_3 = \sigma_\Pi \sigma_\Delta \sigma_\Gamma(P)$ but the three planes through M are not shown to avoid complicating the figure. Hence, $\sigma_M = \sigma_\Pi \sigma_\Delta \sigma_\Gamma$.

Theorem 16.8. *The product of the three reflections in any three mutually perpendicular planes with common point M is the inversion about M.*

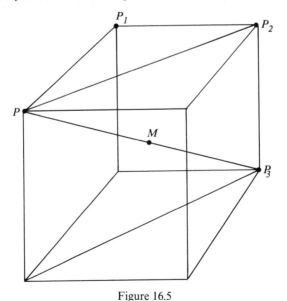

Figure 16.5

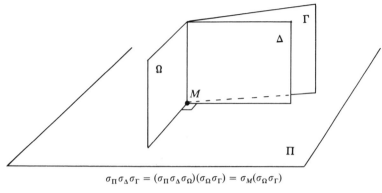

$$\sigma_\Pi \sigma_\Delta \sigma_\Gamma = (\sigma_\Pi \sigma_\Delta \sigma_\Omega)(\sigma_\Omega \sigma_\Gamma) = \sigma_M(\sigma_\Omega \sigma_\Gamma)$$

Figure 16.6

Be sure you can define a rotary reflection. What are the involutory rotary reflections? Suppose plane Π is perpendicular to each of distinct intersecting planes Γ and Δ. Then $\sigma_\Pi \sigma_\Delta \sigma_\Gamma = \sigma_\Delta \sigma_\Gamma \sigma_\Pi$ and $(\sigma_\Pi \sigma_\Delta \sigma_\Gamma)^2 = (\sigma_\Delta \sigma_\Gamma)^2$. Since $(\sigma_\Delta \sigma_\Gamma)^2 = \iota$ iff $\Gamma = \Delta$ or $\Gamma \perp \Delta$, then rotary reflection $\sigma_\Pi \sigma_\Delta \sigma_\Gamma$ is an involution iff $\Gamma \perp \Delta$.

Theorem 16.9. *If Γ and Δ are two intersecting planes perpendicular to plane Π, then rotary reflection $\sigma_\Pi \sigma_\Delta \sigma_\Gamma$ is an involution iff $\Gamma \perp \Delta$.*

Combining the two previous theorems, we have the following determination of the inversions.

Theorem 16.10. *The inversions are the involutory rotary reflections.*

A glance at Figure 16.6 and the equations there explains why a rotary reflection that is not an inversion is sometimes called a *rotary inversion.* Specifically, given distinct intersecting but nonperpendicular planes Γ and Δ each perpendicular to plane Π with M the point common to Γ, Δ, Π, let Ω be the plane through M that is perpendicular to the intersection of Δ and Π. Then the rotary reflection $\sigma_\Pi \sigma_\Delta \sigma_\Gamma$ is equal to $\sigma_M(\sigma_\Omega \sigma_\Gamma)$ with M on the intersection of Γ and Ω by the equations in the figure. Note that we are back to using the method of multiplying by the identity in an appropriately useful form. The method will be used many times below. Now, given point M on the intersection of distinct planes Γ and Ω, we can define planes Δ and Π such that we can read the equations in Figure 16.6 from right to left and obtain the following theorem that is needed later.

Theorem 16.11. *A rotation about a line followed by the inversion about a point on that line is a reflection or a rotary reflection.*

The method of introducing the identity in a useful form and a glance at Figure 16.7 lead to the following theorem.

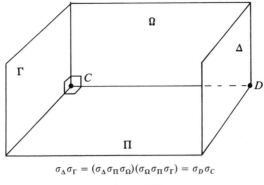

$$\sigma_\Delta \sigma_\Gamma = (\sigma_\Delta \sigma_\Pi \sigma_\Omega)(\sigma_\Omega \sigma_\Pi \sigma_\Gamma) = \sigma_D \sigma_C$$

Figure 16.7

Theorem 16.12. *A product of two inversions is a translation. Conversely, a translation is the product of two inversions, where either the first or the second may be arbitrarily chosen and the other is then uniquely determined.*

Figure 16.8 helps to formulate one of the theorems that deal with "twice the directed distance" or "twice the directed angle." These are left for the exercises however. Instead, we consider the product of three inversions. Given $\sigma_A, \sigma_B, \sigma_C$, by the previous theorem there is a point P such that $\sigma_C \sigma_B = \sigma_P \sigma_A$. So $\sigma_C \sigma_B \sigma_A = \sigma_P = \sigma_P^{-1} = \sigma_A \sigma_B \sigma_C$. If A, B, C are not collinear, then the inversion $\sigma_C \sigma_B \sigma_A$ fixes the vertex D of parallelogram $\square ABCD$. Hence, we must have $\sigma_C \sigma_B \sigma_A = \sigma_D$ since an inversion fixes a unique point.

Theorem 16.13. *A product of three inversions is an inversion. If points A, B, C are not collinear, then $\sigma_C \sigma_B \sigma_A = \sigma_D$ where $\square ABCD$ is a parallelogram. If A, B, C are points, then $\sigma_C \sigma_B \sigma_A = \sigma_A \sigma_B \sigma_C$.*

The proof of the next theorem is left for Exercise 16.1.

Theorem 16.14. *The translations form an abelian group.*

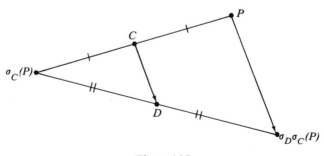

Figure 16.8

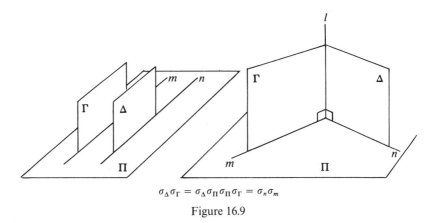

$$\sigma_\Delta \sigma_\Gamma = \sigma_\Delta \sigma_\Pi \sigma_\Pi \sigma_\Gamma = \sigma_n \sigma_m$$

Figure 16.9

A glance at the two parts of Figure 16.9 and the equations there gives the first two parts of the following important theorem.

Theorem 16.15. *A translation is the product of two halfturns about parallel lines, and conversely. A rotation is the product of two halfturns about two intersecting lines, and conversely. A screw is the product of two halfturns about two skew lines, and conversely.*

The third part of the theorem above remains to be proved. Be sure you can define a screw. Given distinct planes Γ and Δ perpendicular to line l and given distinct planes Π and Ω containing line l, let m be the intersection of Γ and Π and let n be the intersection of Δ and Ω. See Figure 16.10. Then

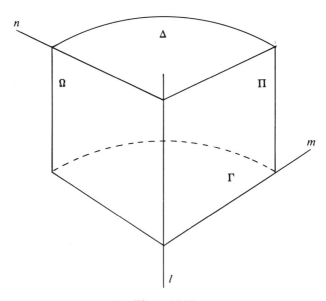

Figure 16.10

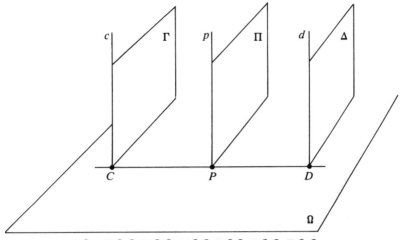

$$\tau_{C,D} = \sigma_\Pi \sigma_\Gamma = \sigma_\Delta \sigma_\Pi = \sigma_p \sigma_c = \sigma_d \sigma_p = \sigma_P \sigma_C = \sigma_D \sigma_P$$

Figure 16.11

m and n are skew and Π and Δ are perpendicular. Hence, $(\sigma_\Omega \sigma_\Pi)(\sigma_\Delta \sigma_\Gamma) = \sigma_\Omega \sigma_\Delta \sigma_\Pi \sigma_\Gamma = \sigma_n \sigma_m$. Conversely, if lines m and n are skew, then let Γ be the plane through m that is parallel to n, let Δ be the plane through n that is parallel to m, let Π be the plane through m that is perpendicular to Δ, and let Ω be the plane through n that is perpendicular to Γ. Then, $\sigma_n \sigma_m = \sigma_\Omega \sigma_\Delta \sigma_\Pi \sigma_\Gamma = (\sigma_\Omega \sigma_\Pi)(\sigma_\Delta \sigma_\Gamma)$, finishing the proof of Theorem 16.15.

The following theorem contains nothing new but does bring several previous results together. See Figure 16.11.

Theorem 16.16. *If point P is the midpoint of $\overline{CD}$ in plane Ω, if lines c, p, d are perpendicular to Ω at C, P, D, respectively, and if planes Γ, Π, Δ are perpendicular to $\overleftrightarrow{CD}$ at C, P, D, respectively, then*

$$\tau_{C,D} = \sigma_\Pi \sigma_\Gamma = \sigma_\Delta \sigma_\Pi = \sigma_p \sigma_c = \sigma_d \sigma_p = \sigma_P \sigma_C = \sigma_D \sigma_P.$$

A *figure* is a nonempty set of points. Let s and t be figures. Then s *is congruent to* t if there is an isometry α such that $\alpha(s) = t$. Transformation α *fixes* s if $\alpha(s) = s$, and transformation α fixes s *pointwise* if $\alpha(P) = P$ for every point P in s. Plane Π is a *plane of symmetry* for figure s if σ_Π fixes s, line l is a *line of symmetry* for figure s if σ_l fixes s, and point P is a *point of symmetry* for figure s if σ_P fixes s.

A figure without a plane of symmetry is said to be *chiral*. This term was introduced by Lord Kelvin in his *Baltimore Lectures*. A figure with a plane of symmetry is then said to be *achiral*. Here we wish only to establish the existence of a chiral figure. In everyday experience, a chiral object is readily at hand. In fact, your right hand is a chiral object. (*Cheir* is Greek for *hand*.) The reflected image of your right hand looks like a left hand. In general, if figure s is chiral and $t = \sigma_\Pi(s)$ for some plane Π, then s and t are called

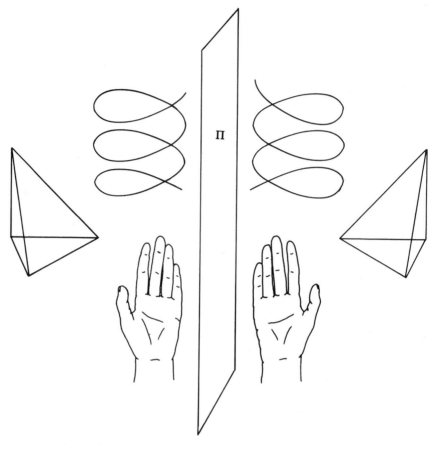

Figure 16.12

enantiomorphs of each other. The helical springs in Figure 16.12 are enantiomorphic, as are the tetrahedra. A left glove is not a right glove but is an
enantiomorph of a right glove. So the image of a right glove under an even
number of reflections is a right glove, and the image of a right glove under an
odd number of reflections is a left glove. Since a left glove (or hand) is not a
right glove (or hand), then it follows that a product of an even number of
reflections is never equal to a product of an odd number of reflections.

Theorem 16.17. *No isometry is both even and odd.*

From this theorem it follows that an even isometry α with fixed point P
is a product $\sigma_\Delta \sigma_\Gamma$ of two reflections in planes through P, since any isometry
with fixed point P is a product of at most three reflections in planes through P.
If $\Delta \parallel \Gamma$, then $\Gamma = \Delta$ and $\alpha = \iota$. Otherwise, Γ and Δ are distinct planes intersecting in some line through P and α is a rotation about that line.

Theorem 16.18. *An even isometry with a fixed point P is either the identity or a rotation about a line through P.*

An odd isometry α with fixed point P is either a reflection or a product of three reflections in planes through P. Then $\sigma_P \alpha$ is even and fixes point P. With $\beta = \sigma_P \alpha$, we have β is either the identity or a rotation about a line through P. Hence, $\alpha = \sigma_P$ or $\alpha = \sigma_P \beta$, and we have the following consequence of Theorem 16.11.

Theorem 16.19. *An odd isometry with a fixed point is either a reflection or a rotary reflection.*

Be sure you can define a glide reflection. Suppose α is an odd isometry with no fixed point. Then α must be a product of three reflections in planes having no point in common. Further, the three factors in the product must be distinct, as their product is not a reflection (Theorem 16.7). Hence, the planes are three distinct planes that are not all parallel and that have no common point. Therefore, the three planes must be perpendicular to some fourth plane. We now have $\alpha = \alpha_\Pi \sigma_\Delta \sigma_\Gamma$ where Γ, Δ, Π are three planes perpendicular to some plane Ω. Since Γ, Δ, Π are neither all parallel nor all pass through some line, then from our knowledge of plane glide reflections (Theorem 8.4) we have the following.

Theorem 16.20. *An odd isometry with no fixed point is a glide reflection.*

The last two theorems give all the odd isometries.

Theorem 16.21. *An odd isometry is a reflection, a glide reflection, or a rotary reflection.*

Suppose isometry α is even and sends points P to point Q with $P \neq Q$. Let Γ be the perpendicular bisector of $\overline{PQ}$. Then $\alpha \sigma_\Gamma$ is odd and fixes Q. So $\alpha \sigma_\Gamma = \sigma_\Omega \sigma_\Pi \sigma_\Delta$ where intersecting planes Δ and Π (possibly equal) are perpendicular to plane Ω. There are planes Δ and Ξ that are perpendicular to Ω such that $\Lambda \perp \Gamma$ and $\sigma_\Pi \sigma_\Delta = \sigma_\Xi \sigma_\Lambda$. See Figure 16.13. Then, $\alpha = (\sigma_\Omega \sigma_\Pi \sigma_\Delta) \sigma_\Gamma = (\sigma_\Omega \sigma_\Xi)(\sigma_\Lambda \sigma_\Gamma)$ and α is the product of two halfturns.

Theorem 16.22. *An even isometry is the product of two halfturns.*

This theorem together with Theorem 16.15 give the following theorem.

Theorem 16.23. *An even isometry is a translation, a rotation, or a screw.*

With one more step, which is left for Exercise 16.3, we have ***The Classification Theorem for Isometries on Space.***

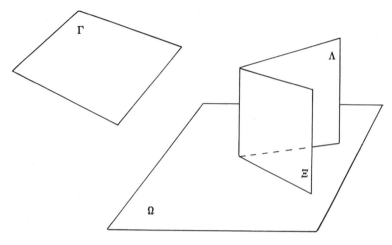

Figure 16.13

Theorem 16.24. *An isometry is exactly one of the following:*

translation,	*reflection,*
rotation,	*glide reflection,*
screw,	*rotary reflection.*

As so many other theorems in mathematics, the theorem above is another that might be called *Euler's Theorem*. (See Section 14.2.) In *A History of Geometrical Methods* by J. L. Coolidge, we find that Euler determined the isometries of space in 1776, while the classification of plane isometries was given in 1831 by Chasles. Michel Chasles (1793–1880) was a mathematical historian and French geometer who made many contributions to the study of projective geometry.

§16.2 Similarities on Space

Our task is clear. We begin with the definitions. If C is a point and $r > 0$, then a **stretch** of ratio r about C is the transformation that fixes C and otherwise sends point P to point P' on $\overrightarrow{CP}$ where $CP' = rCP$. We allow the identity to be a stretch. If $r > 0$, then $\delta_{C,r}$ is the stretch about C of ratio r and $\delta_{C,-r} = \sigma_C \delta_{C,r}$. For $s \neq 0$, we call $\delta_{C,s}$ a **dilation** about C with dilation ratio s. If dilation δ about point P is not an isometry and ρ is a rotation about a line through P, then $\rho\delta$ is a **dilative rotation**. The definitions are chosen such that the rotary reflections are not dilative rotations, since no isometry is a dilative rotation. If $r > 0$, then a **similarity** of ratio r is a transformation α such that $P'Q' = rPQ$ for all points P and Q where $P' = \alpha(P)$ and $Q' = \alpha(Q)$. If similarity

α is a product of a stretch and an even isometry, then α is **direct**; if similarity α is a product of a stretch and an odd isometry then α is **opposite**

Let's put down some things that follow rather quickly from the definitions. First, the similarities obviously form a group. Then, given a stretch $\delta_{C,r}$ and any line l, the restriction of $\delta_{C,r}$ to a plane containing C and l is just a plane stretch about C of dilation ratio r. Hence, $\delta_{C,r}(l)$ is a line parallel to l. It follows that a stretch is a collineation and a similarity. If α is a similarity of ratio r and P is any point, then $\alpha\delta_{P,1/r}$ is a similarity of ratio 1 and, hence, is some isometry β. So $\alpha = \beta\delta_{P,r}$ and a similarity is just a stretch about P followed by an isometry. The rest of the following theorem then follows from the fact that a stretch changes the size of a glove but not the handedness of the glove.

Theorem 16.25. *The similarities form a group. A stretch is a collineation and a similarity. If α is any similarity of ratio r and P is any point, then there is an isometry β such that $\alpha = \beta\delta_{P,r}$. A similarity is a collineation that is either direct or opposite but not both. The composite of two direct similarities or of two opposite similarities is direct; the composite of a direct similarity and an opposite similarity (in either order) is opposite. The direct similarities form a group.*

We want to show that a similarity that is not an isometry has a unique fixed point. The proof uses the plane analogue, Theorem 13.9. Suppose similarity α has ratio r with $r \neq 1$. Then α is not an isometry and cannot have two fixed points. We want to show α has at least one fixed point. Suppose $\alpha(A) = A' \neq A$. Let $s = +r$ if α is direct; let $s = -r$ if α is opposite. Let Q be the point on $\overleftrightarrow{AA'}$ such that $QA'/QA = s$. Let $\delta = \delta_{Q,s}$. Then $\delta(A) = A'$, δ has similarity ratio r, and α and δ are either both direct or both opposite. Hence, $\alpha\delta^{-1}$ is an even isometry that fixes point A'. If $\alpha\delta^{-1}$ is the identity, then α is a dilation and fixes point Q. If $\alpha\delta^{-1}$ is not the identity, then $\alpha\delta^{-1}$ is a rotation ρ about some line l with A' on l (Theorem 16.18). In this case, let Γ be the plane through Q that is perpendicular to l. See Figure 16.14. Since both δ and ρ fix plane Γ, then $\alpha(\Gamma) = \rho\delta(\Gamma) = \Gamma$. The restriction of α to the plane Γ induces a plane similarity on Γ that is not an isometry. So (Theorem 13.9), there is some point S on Γ such that $\alpha(S) = S$.

Theorem 16.26. *A similarity without a fixed point is an isometry.*

Suppose α is a similarity with fixed point S and similarity ratio r. Let $s = +r$ if α is direct; let $s = -r$ if α is opposite. Let $\delta = \delta_{S,s}$. Then $\alpha\delta^{-1}$ is an even isometry that fixes point S. So either $\alpha = \delta$ or else $\alpha\delta^{-1} = \rho$ where ρ is a rotation about some line through S. In the second case, $\alpha = \rho\delta$ and α is a dilative rotation when $r \neq 1$. This gives **The Classification Theorem for Similarities on Space**.

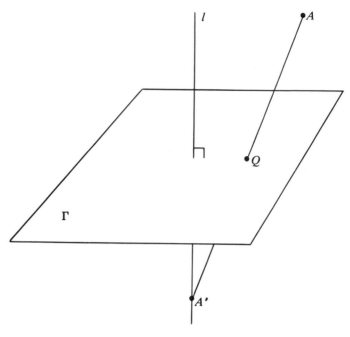

Figure 16.14

Theorem 16.27. *A similarity that is not an isometry is either a dilation or a dilative rotation.*

Thus the similarities of space are the isometries, the dilations, and the *spirals*, as the dilative rotations are sometimes called.

§16.3 Exercises

16.1. Prove: The translations form an abelian group (Theorem 16.14).

16.2. Show that a screw is neither a translation nor a rotation.

16.3. Verify the "exactly" in Theorem 16.24.

16.4. Prove the last statement in Theorem 16.7.

16.5. Prove or disprove: A similarity that is not an isometry is a dilation, a *stretch rotation* (a nonidentity stretch about some point followed by a rotation about a line through that point), or a *stretch reflection* (a nonidentity stretch about some point followed by a reflection in a plane through that point).

16.6. True or False
(a) Dilation $\delta_{P,\,-2}$ has dilation ratio -2 and similarity ratio $+2$.
(b) Only ι is both a translation and a rotation.

   (c) A product of four reflections is a product of two reflections.

   (d) An isometry with three fixed points is a reflection.

   (e) A product of two rotations might be a screw.

   (f) A product of two screws might be a rotation.

   (g) A screw followed by an inversion is another screw.

   (h) There is always a plane perpendicular to each of three given planes with no common point.

   (i) There is a line perpendicular to each of three given planes with no common point.

   (j) A similarity that is not an isometry has a fixed point.

16.7. Write out a proof of Theorem 16.11.

16.8. Which isometries are involutions? Which similarities are involutions?

16.9. Give the Cayley table for $\langle \sigma_P, \sigma_\Pi \rangle$ where point P is on plane Π.

16.10. A *dilatation* is a collineation α such that $l \parallel \alpha(l)$ for every line l. Which similarities are dilatations?

16.11. Describe $\sigma_\Pi \sigma_P$ when point P is off plane Π.

16.12. If α is a similarity, τ is a translation, ρ is a rotation, and η is a screw, then show $\alpha\tau\alpha^{-1}$ is a translation, $\alpha\rho\alpha^{-1}$ is a rotation, and $\alpha\eta\alpha^{-1}$ is a screw.

16.13. Prove or disprove: Isometry $\sigma_\Gamma \sigma_l$ is a glide reflection iff line l is parallel to plane Γ but off Γ.

16.14. Prove or disprove: Isometry $\sigma_l \sigma_P$ is a glide reflection iff point P is off line l.

16.15. Prove or disprove: A rotation about line l fixes plane Π iff $l \perp \Pi$.

16.16. List several conjectures about isometries that involve "twice the directed distance" or "twice a directed angle."

16.17. Prove or disprove: Every isometry is a product of isometries of order 4.

16.18. Why does a mirror interchange *right* and *left* but not *top* and *bottom*?

16.19. Show that the product of two dilations is a dilation or a translation.

16.20. Find and prove a necessary and sufficient condition for translation τ and rotation ρ to commute.

16.21. Describe the even symmetries of a regular tetrahedron.

16.22. Describe the even symmetries of a cube.

16.23. Which similarities comute with a given screw?

16.24. Find and verify the equations for the inversion about (a, b, c) in Cartesian three-space.

16.25. If line l passes through the two points $(0, 0, 0)$ and (a, b, c), then what are the equations for σ_l?

Chapter 17

Space and Symmetry

§17.1 The Platonic Solids

If a convex polyhedron has v vertices, e edges, and f faces, then $v - e + f = 2$. This simple but elegant theorem is known as Euler's Formula, even though Descartes had stated the equation over 100 years before Euler gave the result in 1752. A polyhedron is **convex** if all the vertices not on any given face lie on one side of the plane containing that face. To prove the famous formula, imagine that all the edges of a convex polyhedron are dikes, exactly one face contains the raging sea, and all other faces are dry. We break dikes one at a time until all the faces are inundated, following the rule that a dike is broken only if this results in the flooding of a face. Now, after this rampage, we have flooded $f - 1$ faces and so destroyed exactly $f - 1$ dikes. Noticing that we can walk with dry feet along the remaining dikes from any vertex P to any other vertex along exactly one path, we conclude there is a one-to-one correspondence between the remaining dikes and the vertices excluding P. Hence, there remain exactly $v - 1$ unbroken dikes. So $e = (f - 1) + (v - 1)$ and we have proved **Euler's Formula**.

Theorem 17.1. *If a convex polyhedron has v vertices, e edges, and f faces, then*

$$v - e + f = 2.$$

Undoubtedly you have heard there are exactly five regular solids. These are illustrated in Figure 17.1 and, except for the cube, are named after their number of faces. A convex solid and the polyhedron that is the surface of that solid have the same name and serve the same function for our purposes. The regular tetrahedron, the cube, and the regular octahedron occur naturally in

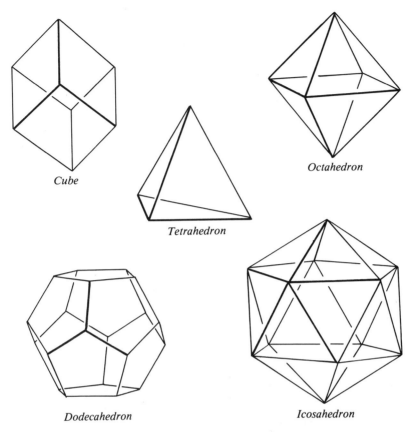

Cube

Tetrahedron

Octahedron

Dodecahedron

Icosahedron

Figure 17.1

common crystals. The regular dodecahedron and the regular icosahedron cannot occur as crystals. A toy regular dodecahedron has been found near Padua in Etruscan ruins from 500 B.C. All the regular solids have been observed as the shapes of certain microscopic marine animals called radiolarians. The abstract concept of a regular solid is due to Theatetus (419–369 B.C.), who is also famous for laying the foundation of the study of irrationals that appears in Euclid's *Elements*. Theatetus is given credit for the determination that there are exactly five regular solids. However, the solids are usually named after his student, colleague, friend, and founder of the Academy at Athens, Plato.

The symmetry groups of the Platonic solids play an important role in the determination of all the finite groups of isometries. All such groups will be found in the next section. As we verify below that there are exactly five regular polyhedra, we shall also learn a few things about the symmetry groups of these famous polyhedra.

For our purposes, a polyhedron is **regular** if the polyhedron is convex, all its faces are regular *p*-gons for some fixed *p*, and the same number *q* of these

Table 17.1

p	q	v	e	f	Platonic Solid
3	3	4	6	4	Regular Tetrahedron
3	4	6	12	8	Regular Octahedron
3	5	12	30	20	Regular Icosahedron
4	3	8	12	6	Cube
5	3	20	30	12	Regular Dodecahedron

faces meet at each vertex. So $p > 2$ and $q > 2$. We suppose we have such a regular polyhedron with v vertices, e edges, and f faces. Since each edge is shared by two faces and also contains two vertices, then pf and qv both give twice the number of edges. So, $qv = 2e = pf$, and, by Euler's Formula,

$$2 = v - e + f = v - (qv)/2 + (qv)/p$$
$$= [2p + 2q - pq](v/2p)$$
$$= [4 - (p - 2)(q - 2)](v/2p).$$

Therefore, we have

$$v = \frac{4p}{2p + 2q - pq}, \qquad e = \frac{qv}{2}, \qquad f = \frac{qv}{p}, \quad \text{and} \quad (p - 2)(q - 2) < 4.$$

These first three equations tell us that v, e, f are determined once p and q are known. It follows that a given p and q determine at most one regular solid. Since p and q are integers greater than 2, then the inequality is very easily solved. There are exactly five solutions for p and q to the inequality, and these are given in the first two columns of Table 17.1, along with corresponding values for v, e, f.

We have shown there are at most five regular polyhedra. These are given by the rows of Table 17.1. We also need to argue that there exist polyhedra satisfying the conditions demanded by these solutions p and q. Do twelve congruent regular pentagons really fit together as in the figure to form the regular dodecahedron or do they miss by a small amount not discernible in the picture? A similar question might be posed for the other proposed polyhedra. The easiest way out of this problem is to produce the polyhedra in coordinate geometry. By the distance formula, the points

$$(1, 1, 1), \qquad (1, -1, -1), \qquad (-1, 1, -1), \qquad (-1, -1, 1)$$

are seen to be the four vertices of a regular tetrahedron with edges of length $\sqrt{2}$, and points

$$(\pm 1, \pm 1, \pm 1)$$

are seen to be the eight vertices of a cube. See Figure 17.2a. Since we probably did not doubt the existence of these particular two polyhedra, it is interesting

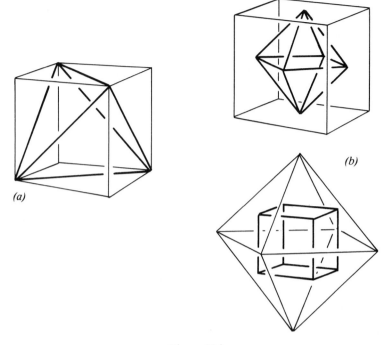

Figure 17.2

to note that the vertices of the tetrahedron are among those of the cube and that every symmetry of the tetrahedron is also a symmetry of the cube. Thus, the symmetry group of a cube contains the symmetry group of a tetrahedron. The polyhedron formed by joining the centers of adjacent faces of the cube has the six vertices

$$(\pm 1, 0, 0), \qquad (0, \pm 1, 0), \qquad (0, 0, \pm 1)$$

and is easily seen to be a regular octahedron. See Figure 17.2b. So a regular octahedron does exist and every symmetry of this octahedron is a symmetry of the cube. A polyhedron with edges obtained by joining the centers of adjacent faces of a given regular polyhedron is called the **dual** of the given polyhedron. The dual of a tetrahedron is a tetrahedron. However, not only is the regular octahedron the dual of the cube, but the dual of the regular octahedron is another cube. See Figure 17.2b again. This means that the cube and the regular octahedron have the same symmetry group.

Looking at Figure 17.1 again, we see that if either the regular icosahedron or the regular dodecahedron exists then both must exist as each would have the other as a dual. Further, if such polyhedra do exist, then it follows that they must have the same symmetry group. Let g be the positive number such that $g^2 = g + 1$. We claim the points

$$(\pm 1, 0, \pm g), \qquad (0, \pm g, \pm 1), \qquad (\pm g, \pm 1, 0)$$

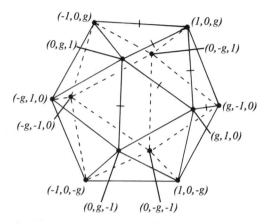

Figure 17.3

are the twelve vertices of a regular icosahedron with edge of length 2. See
Figure 17.3. Because this set of twelve vertices is symmetric with respect to
each of the three planes containing two of the coordinate axes, we need check
only the lengths of the six edges indicated by a bar in the figure. This is easily
done using the distance formula. The vertices of the dual of the icosahedron
above have coordinates that, when multiplied by $3/g$ to avoid fractions, give
the points

$$(0, \pm 1, \pm g^2), \qquad (\pm g^2, 0, \pm 1), \qquad (\pm 1, \pm g^2, 0), \qquad (\pm g, \pm g, \pm g)$$

as the twenty vertices of a regular dodecahedron. (Note $2g + 1 = g^2 g$ since
$g^2 = g + 1$.) See Figure 17.4. We cannot help but notice that the vertices
$(\pm g, \pm g, \pm g)$ of a cube are among those of the regular dodecahedron and
that every symmetry of this cube must be a symmetry of the regular dodecahedron and, hence, of the regular icosahedron that is the dual of the dodecahedron.

Let's recapitulate. There are exactly five regular polyhedra; these are the
surfaces of the Platonic solids of Table 17.1. The vertices of a regular tetrahedron are among those of a cube, and so the symmetry group of the cube

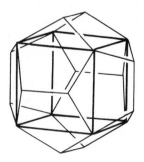

Figure 17.4

contains the symmetry group of the regular tetrahedron. Since cubes and regular octahedra are dual to each other, the cube and the regular octahedron have the same symmetry group. The vertices of a cube are among those of a regular dodecahedron, and so the symmetry group of the regular dodecahedron contains the symmetry group of the cube. Since the regular dodecahedron and the regular icosahedron are dual to each other, then the regular dodecahedron and the regular icosahedron have the same symmetry group. Our results are summarized in **_Theatetus' Theorem_**.

Theorem 17.2. *There are exactly five regular polyhedra. The symmetry group of a cube is the symmetry group of a regular octahedron. The symmetry group of a regular dodecahedron is the symmetry group of a regular icosahedron. The symmetry group of a regular icosahedron contains the symmetry group of a regular octahedron, which, in turn, contains the symmetry group of a regular tetrahedron.*

Each of the Platonic solids can be inscribed in a sphere. Any vertex V of a Platonic solid can be brought to different vertex V' by a symmetry of the sphere that is a rotation about the axis of the sphere that is perpendicular to a plane containing V, V', and the center of the sphere. Each of the Platonic solids also has a plane of symmetry, and so the full group of symmetries of each Platonic solid contains odd isometries. However, we wish to introduce at this time only the subgroups of even isometries for the Platonic solids. The group of all even isometries that fix a regular tetrahedron is called a **_tetrahedral group_** and is denoted by T; the group of all even isometries that fix a regular octahedron (or cube) is called an **_octahedral group_** and is denoted by O; the group of all even isometries that fix a regular icosahedron (or regular dodecahedron) is called an **_icosahedral group_** and is denoted by I; and by the last part of Theatetus' Theorem we may suppose that I contains O and that O contains T. Again, note that T, O, and I are not the symmetry groups of the Platonic solids since these groups contain no odd isometries. With q and v as in Table 17.1, we can compute the orders of these three groups at once. Let V_1 and V_2 be adjacent vertices of one of the Platonic solids. Once the images of V_1 and V_2 are determined then the images of all vertices are determined, as the identity is the only *even* isometry that fixes each of two adjacent vertices. By a rotation or the identity, vertex V_1 can be made to coincide with any of the v vertices of the solid. However, V_2 must then go to one of the q vertices that are adjacent to that one and this too can be accomplished by a rotation or the identity. Since the images of V_1 and V_2 determine the images of all vertices, we see that the group of all even isometries fixing the solid has order vq.

Theorem 17.3. *Groups T, O, and I have orders 12, 24, and 60, respectively. Group I contains group O, and group O contains group T.*

Two vertices, two edges, or two faces of a polyhedron are said to be **_equivalent_** if there is a symmetry of the polyhedron that takes one onto the

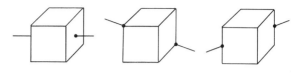

Figure 17.5

other. The argument above showed that all the vertices and all the edges of a Platonic solid are respectively equivalent. Since each edge of a Platonic solid lies on a plane of symmetry, it follows that all the faces of a Platonic solid are also equivalent. Conversely, if all the vertices, all the edges, and all the faces of a polyhedron are respectively equivalent, then it can be shown that each vertex is surrounded by the same number of regular polygonal faces. The following theorem, which is not used later, then gives an alternative character- ization for the Platonic solids.

Theorem 17.4. *A convex polyhedron is regular iff the polyhedron has equivalent vertices, equivalent edges, and equivalent faces.*

Let's describe the even symmetries of a cube. Figure 17.5 will help. First, about each of the three axes that contain centers of opposite faces there are rotations of 90°, 180°, and 270°. This gives nine rotations. Secondly, about each of the four diagonals there are rotations of 120° and 240°. This gives eight more rotations. (Many people are surprised to realize that a cube has a three-fold symmetry. This is probably because we seldom visualize the cube from the perspective given in Figure 17.1.) Also, there are halfturns about each of the six lines through the midpoints of opposite edges. This gives six addi- tional rotations. Together with the identity we now have a total of twenty- four even isometries and should stop looking for more (Theorem 17.3).

Three-dimensional models are almost a necessity for studying many of the more complicated polyhedra. The paperback *Mathematical Models* by H. M. Cundy and A. P. Rollett is an excellent guide for building such models. We cannot resist describing the construction of a model of a regular dode- cahedron that was given by the eminent Polish mathematician Hugo Steinhaus in his classic *Mathematical Snapshots*. Cut two enlarged copies of Figure 17.6 from poster board. (Use a protractor to help construct the pentagons.) Crease

Figure 17.6

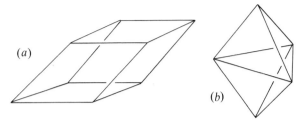

Figure 17.7

each copy along the edges of the inner pentagon. Place the two copies flat against each other with the creases outside and with one copy rotated 36° with respect to the other. Then, weave a rubber band alternately above and beneath the corners of the double star, holding the double star flat with one hand. A perfect model of the regular dodecahedron will pop up when you lift your hand if the rubber band is just right. Try it.

Although we have no intention of making a thorough study of polyhedra, we shall now look at several more polyhedra in order to provide examples and exercises for the discussion on finite symmetry groups. All six faces of the *rhombohedron* in Figure 17.7a are rhombi and all these faces are congruent. Copies of this polyhedron can be stacked together to fill all of space without leaving any gaps. This is what we mean when we say a polyhedron **tessellates** space. Any polygon can be the base of a pyramid. A *bipyramid* is a pyramid together with its image under the reflection in the plane containing its base. The bipyramid in Figure 17.7b is formed by two regular tetrahedra. For this bipyramid, it is easy to see that the faces are equivalent. Therefore, the Platonic solids are not the only polyhedra having all faces regular polygons and all faces congruent.

Among the pyramids with a triangular base, the most interesting are the *disphenoids*. See Figure 17.8. These are formed by folding an acute triangle along the segments joining the midpoints of the sides to have the vertices of the acute triangle coincide at the apex of the pyramid. Obviously, the four faces of a disphenoid are congruent. Models can be quickly made with paper,

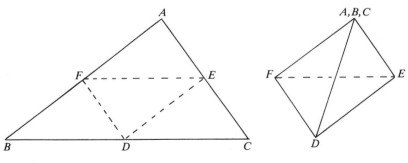

Figure 17.8

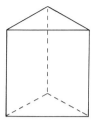 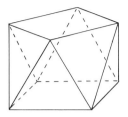

Figure 17.9

scissors, and Scotch tape. You should make two such models: a chiral model based on a scalene triangle and an achiral model based on an isosceles triangle. With these models it is easy to see that the faces of a disphenoid are equivalent and the vertices of a disphenoid are equivalent. The edges of a disphenoid are equivalent iff the original triangle is equilateral, in which case the disphenoid is a Platonic solid.

Any polygon can be the base of a prism. The sides of a prism are rectangles; by proper selection of the height so that the rectangles are squares, a prism can have faces that are all regular polygons. In case the base is also a square, the prism is a cube. Note that a prism with regular n-gon as base has a point of symmetry iff n is even. The vertices of a prism with a regular polygon as base are equivalent. An *antiprism* is formed by twisting the top of a prism with regular n-gon as base through an angle of $(180/n)°$ and forming new sides that are triangular. By proper selection of the height so that the sides are all equilateral triangles, an antiprism can have faces that are all regular polygons. In case the base is also an equilateral triangle, the antiprism is a regular octahedron. (Hold Figure 17.1 at an angle.) In Figure 17.9 the antiprism with a square base has triangular sides that are equilateral. Note that an antiprism with regular n-gon as base has a point of symmetry iff n is odd. The vertices of an antiprism with a regular polygon as base are equivalent. Prisms and antiprisms will be used extensively in the next section.

In addition to the five Platonic solids and certain obvious infinite families of prisms and antiprisms just described, it can be shown that there are exactly thirteen convex polyhedra having equivalent vertices and congruent edges. These are called the *Archimedean solids*. The faces of an Archimedean solid are regular polygons. These solids may be denoted by giving the number of the sides of the faces in cyclic order about any one vertex: $3 \cdot 6^2$, $3 \cdot 8^2$, $4 \cdot 6^2$, $(3 \cdot 4)^2$, $4 \cdot 6 \cdot 8$, $3 \cdot 4^3$, $3^3 \cdot 4$, $3 \cdot 10^2$, $(3 \cdot 5)^2$, $5 \cdot 6^2$, $4 \cdot 6 \cdot 10$, $3 \cdot 4 \cdot 5 \cdot 4$, and $3^4 \cdot 5$. For example, the *truncated octahedron* $4 \cdot 6^2$ has a square and two regular hexagons surrounding each vertex. If we truncate each vertex of a regular octahedron so that the original triangular faces become regular hexagons (Figure 17.10a), the resulting solid is the truncated octahedron (Figure 17.10b). Clearly the edges of the truncated octahedron are not all equivalent. One of the most interesting things about the truncated octahedron is that the polyhedron tessellates space.

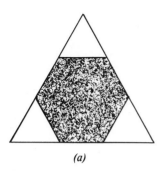

(a)

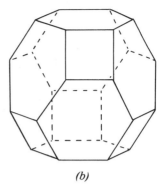

(b)

Figure 17.10

The **vertex figure** at vertex A of a polygon is the segment joining the midpoints of the two sides of the polygon that contain A; the **vertex figure** at vertex A of a polyhedron is the union of the vertex figures at A of all the faces of the polyhedron that contain A. Although in general a vertex figure of a polyhedron will not lie in one plane, vertex figures are more convenient to work with than solid angles. The vertex figure at each vertex of a cube is a triangle. See Figure 17.11a. The polyhedron whose set of edges is the union of the eight vertex figures of a cube is a *cuboctahedron*. This is the Archimedean solid $(3 \cdot 4)^2$, where each vertex is surrounded in cyclic order by a triangle, a square, a second triangle, and a second square. The convex solid with vertices

$$(0, \pm 1, \pm 1), \qquad (\pm 1, 0, \pm 1), \qquad (\pm 1, \pm 1, 0)$$

is a cuboctahedron. This Archimedean solid not only has congruent edges but equivalent edges as well. Our next polyhedron is the "Archimedean dual" of

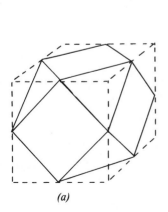

(a)

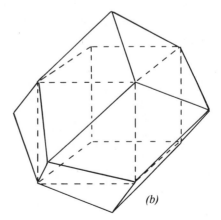

(b)

Figure 17.11

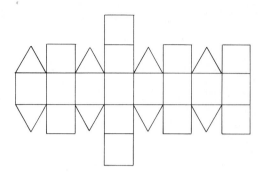

Figure 17.12

the cuboctahedron. (We shall not pause to define the concept of the Archimedean dual of an Archimedean solid.) The easiest approach to this non-Archimedean solid is to first visualize space tessellated with cubes that are alternately colored red and black—think of an infinite three-dimensions checker board. Now, each of the red cubes consists of six square-based pyramids having the center of the red cube as common vertex and one face of the red cube as base. The solid formed by a black cube together with the six adjacent red pyramids, one from each of six different adjacent red cubes, is called a *rhombic dodecahedron*. See Figure 17.11b. The twelve faces of the rhombic dodecahedron are rhombi. Further, these faces are equivalent. From our method of construction, it is clear that the rhombic dodecahedron tessellates space.

Folding the templet in Figure 17.12 into a convex polyhedron, we obtain the *rhombicuboctahedron* in Figure 17.13a. The rhombicuboctahedron is the Archimedean solid $3 \cdot 4^3$. Each vertex is surrounded by an equilateral triangle and three squares. Now, giving the roof of the rhombicuboctahedron a twist of 45°, we have *Sommerville's solid*. See Figure 17.13b. Each vertex of Sommerville's solid is also surrounded by an equilateral triangle and three squares. Sommerville's solid, which is also called the pseudo-rhombicuboctahedron or

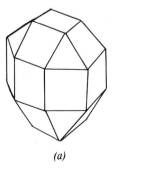

(a)

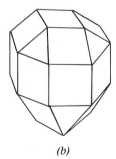

(b)

Figure 17.13

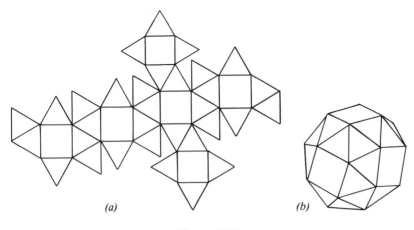

(a)   (b)

Figure 17.14

the elongated square gyrobicupola, was discovered by the English mathematician, D. M. Y. Sommerville in 1904. Sommerville's solid is not an Archimedean solid, although all of its faces are regular polygons and all the vertex figures are congruent.

The Archimedean solid $3^4 \cdot 4$ is called the *snub cube* and the Archimedean solid $3^4 \cdot 5$ is called the *snub dodecahedron*. These are perhaps the most interesting of the Archimedean solids because each is chiral. The templet in Figure 17.14a can be folded to make the snub cube shown in Figure 17.14b. By folding the templet "the other way," which in essence turns the templet over, we obtain the enantiomorph of the snub cube in Figure 17.14b. In Figure 17.15a, this enantiomorph is shown with a cube in position such that each of the six faces of the snub cube lies in a face of the cube. The coplanar square faces of the cube and snub cube have the same center but their edges are not parallel. The Archimedean solids are studied in *Solid Geometry* by L. Lines, where it is shown that the relative position of the coplanar squares is given in Figure 17.15b, with the cube having edge of length 2, $c^2 = 2b - 1$,

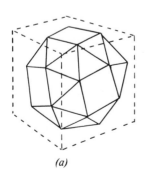
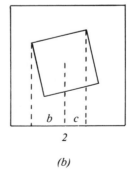

(a)   (b)

Figure 17.15

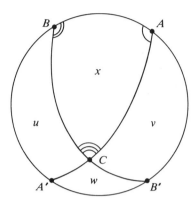

Figure 17.16

and b is the real root of the equation $x^3 + x^2 + x - 1 = 0$. (An approximation for b is 0.544.) A regular dodecahedron can be placed relative to a given snub dodecahedron such that each of the twelve pentagonal faces of the snub dodecahedron lies in a face of the regular dodecahedron.

The fact that the sum of the measures of the angles of a spherical triangle is greater than 180 is needed in the next section. We shall finish this section by proving this lemma. A spherical triangle has sides that are three minor arcs of great circles on a sphere. Recall that a great circle is the intersection of the sphere with a plane through the center of the sphere. We suppose we have a spherical triangle with vertices A, B, C. Let A' and B' be the antipodal points corresponding to A and B respectively. The great circle containing A and B passes through A' and B' and divides the sphere into two hemispheres. Let x, u, v, w be the areas in the hemisphere containing C of the regions indicated in Figure 17.16. The hemisphere not containing C has regions that are congruent to those we see in the figure. Let s be the area of the sphere. (We needn't know $s = 4\pi r^2$.) Then $m \angle A/360 = (x + u)/s$. This gives the first of the similarly derived equations

$$x + u = (m\angle A)(s/360),$$

$$x + v = (m\angle B)(s/360),$$

$$x + w = (m\angle C)(s/360).$$

Since $x + u + v + w = s/2$, adding the three equations above, we get the single equation

$$2x = (m\angle A + m\angle B + m\angle C - 180)(s/360).$$

Since $x > 0$, we have the desired result

$$m\angle A + m\angle B + m\angle C > 180.$$

Theorem 17.5. *The sum of the measures of the angles of a spherical triangle is greater than 180.*

§17.2 Finite Symmetry Groups on Space

The determination of the finite groups of isometries on space was accomplished in 1830 by the mineralogist Johann Friedrich Christian Hessel (1796–1872). This work received no recognition among his contemporaries and remained almost unknown until republished in 1897. The same determination was made independently, more elegantly, but later by Auguste Bravais (1811–1863), who made notable contributions to botany among his many scientific accomplishments. This section is devoted to finding all the finite groups of isometries. We begin by looking for all the finite groups that contain only even isometries. Since each of a nonidentity translation and a screw has infinite order, such a group can contain only rotations and the identity. Any group containing only rotations and the identity is called a ***rotation group***. For example, all the rotations with axes concurrent at point C together with the identity form a rotation group. This follows from the fact that the product of two such rotations and the inverse of such a rotation are even and fix C and the fact that the only even isometries with a fixed point C are the identity and the rotations with axes through C. At the other extreme, the group consisting of only the identity is also called a rotation group.

A finite isometry group G is sometimes called *point group* because, as we shall see, there is always a point C fixed by every member of G. So the axes of all rotations in G are then concurrent at C. In order to prove the existence of such a point C, we first assume G is a rotation group containing rotations α and β with distinct parallel axes m and n, and we seek a contradiction. (We know $\beta\alpha$ is a rotation or a translation from our results on plane isometries.) Let Π be the plane containing m and n. Then there are planes Γ and Δ such that $\alpha = \sigma_\Pi\sigma_\Gamma$ and $\beta = \sigma_\Delta\sigma_\Pi$. See Figure 17.17a. Since Γ, Δ, Π have no common point, then $\sigma_\Pi\sigma_\Delta\sigma_\Gamma$ is a glide reflection. Thus $(\sigma_\Pi\sigma_\Delta\sigma_\Gamma)^2$ is a nonidentity translation and consequently not in G. However, if α and β are both in group G, then $\beta^{-1}\alpha^{-1}\beta\alpha$ must be in G. The desired contradiction follows from the equation $\beta^{-1}\alpha^{-1}\beta\alpha = (\sigma_\Pi\sigma_\Delta\sigma_\Gamma)^2$. Seeking another contradiction

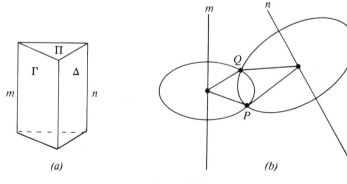

Figure 17.17

we now assume rotation group G contains rotations α and β with axes m and n, respectively, that are skew. Supposed rotation $\beta\alpha$ must fix some point P, which is off both m and n. With $Q = \alpha(P)$, the axis m of α is on the perpendicular bisector of $\overline{PQ}$. See Figure 17.17b. Since $\beta(Q) = P$, then the axis n of β is on the perpendicular bisector of $\overline{PQ}$. Hence, we have the desired contradiction that m and n are coplanar. The argument shows that the product of rotations α and β with skew axes m and n, respectively, cannot be a rotation. In passing, it is not too difficult to see that $\beta\alpha$ cannot be a translation either. Since a rotation fixes its axis, then $\beta\alpha$ cannot be a translation unless $m \parallel \beta(m)$ and $\alpha^{-1}(n) \parallel n$, which requires that m and n be perpendicular and that α and β are both halfturns. However, even in this case $\beta\alpha$ is a screw where the rotation is a halfturn about the common perpendicular to m and n. Somewhat by serendipity, we have the following.

Theorem 17.6. *If rotations α and β have skew axes, then $\beta\alpha$ is a screw.*

We have shown two axes of the rotations in a rotation group can be neither parallel nor skew. It may seem that we have finished the proof that all axes are concurrent. We have not. What we have shown is that any *two* axes intersect. It is necessary to show that any *three* axes are concurrent to be sure that all the axes are concurrent. However, if α, β, γ are three rotations with axes that intersect in pairs but are not concurrent, then $\gamma\beta\alpha$ is a screw. The argument for this (Exercise 17.2) follows from Theorem 17.6 and Figure 17.18. We have then proved the following.

Theorem 17.7. *The axes of the rotations in a rotation group are concurrent.*

Since $\alpha\sigma_\Delta\sigma_\Gamma\alpha^{-1} = \alpha\sigma_\Delta\alpha^{-1}\alpha\sigma_\Gamma\alpha^{-1} = \sigma_{\alpha(\Delta)}\sigma_{\alpha(\Gamma)}$ for any planes Γ and Δ and any isometry α, then it follows that the conjugate $\alpha\rho\alpha^{-1}$ of a rotation ρ of order n with axis l is a rotation of order n with axis $\alpha(l)$. In particular, since α and ρ in a group implies $\alpha\rho\alpha^{-1}$ is in the group, then the images of an axis of rotation of order n under the elements of a symmetry group are axes of rotation of order n in the group. This is important enough to mention as a theorem.

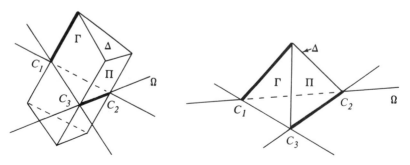

Figure 17.18

Theorem 17.8. *If* α *and* ρ *are in isometry group* G *where* ρ *is a rotation of order* n *with axis* l, *then* G *contains a rotation of order* n *with axis* $\alpha(l)$.

Since the axes of the rotations in a rotation group H are concurrent, there is a point fixed by all the rotations in H. If there are rotations in H with distinct axes, then this point is uniquely determined as the point of concurrency of all the axes of the rotations. If all the rotations have the same axis l, then every point on l is fixed by all the rotations in H. If H contains only the identity, then every point is fixed. Group G *fixes* point C if every isometry in G fixes C. So a rotation group fixes a point. We wish to show below that every finite isometry group G has a fixed point. By the remarks above, we need consider only finite groups G containing odd isometries.

Suppose finite isometry group G contains an odd isometry β. Half the elements in G are then even isometries and half the elements in G are odd isometries. Since G is finite, the even isometries form a rotation subgroup H. The odd isometries are the composites of the elements in H followed by β. Since a glide reflection has infinite order, then odd isometry β must be a reflection or a rotary reflection. Since a reflection fixes all the points on a plane and a rotary reflection has a unique fixed point, if $H = \{\iota\}$, then $G = \{\iota, \beta\}$ and G has a fixed point. Suppose next that H has rotations and all the rotations have the same axis l. If $\beta = \sigma_\Pi$, then Π must intersect l at some point C as otherwise l and $\sigma_\Pi(l)$ would be parallel axes of rotation in finite rotation group H, a contradiction. If β is a rotary reflection, then β must fix l, the only axis of rotation, and so the fixed point C of β must lie on l. In either case, since C is fixed by β and by each element in H, then C is fixed by each element in G. Finally, suppose that H has rotations with distinct axes concurrent at point C. Since β permutes the axes of rotation, then β must fix the point of concurrency of these axes. Again, since β and the elements in H fix C, then all the elements in G fix C.

Theorem 17.9. *A rotation group has a fixed point. A finite group of isometries has a fixed point.*

The finite groups of plane isometries can be interpreted as coming from finite rotation groups. A nonidentity plane rotation about point P in a given plane can be interpreted as the restriction of a rotation (on space) with axis perpendicular to the given plane at P. With this in mind we now let C_n denote the cyclic group generated by a rotation of order n and define C_1 to be the group containing only the identity. A plane reflection with axis l in a given plane can be interpreted as the restriction to the given plane of the halfturn (on space) about line l. That is, the rotation of $180°$ about line l has the same effect on a plane Π through l as the plane reflection in l on Π. With this in mind, we now let D_n denote the group generated by a rotation ρ of order n and a halfturn σ when the axes of ρ and σ are perpendicular. Group D_n has order $2n$ and contains n halfturns about axes that are perpendicular to the axis of ρ

at a common point and that form angles of $(360/2n)°$ with each other. Rotation group C_n is called a *cyclic group*, and notation group D_n is called a *dihedral group*, where the latter term designates any group generated by distinct elements σ and ρ where ρ has finite order and both σ and $\rho\sigma$ are involutions.

Since the identity is not a rotation, you might have noticed the dihedral group of two elements from the plane isometries did not get carried up to a group D_1 on space. Such a designation would have been superfluous anyway since the group would contain exactly the identity and one halfturn and, hence, would be a group C_2. In classifying groups, if lines l and m are distinct, we identify both groups $\{\iota, \sigma_l\}$ and $\{\iota, \sigma_m\}$ because there is an isometry taking l to m. To be very formal, a geometer identifies groups G and H if the elements of one are the conjugates of the elements of the other by some fixed isometry. Of course, groups G and H are called *conjugate groups* in this case. A geometer does not identify $\{\iota, \sigma_l\}$ and $\{\iota, \sigma_P\}$ just because each is generated by an involution. Such an identification is of interest to those studying abstract algebra but not to us.

Groups C_n and D_n are rotation groups and so contain only even isometries. This phenomenon may surprise you. You should be able to see that a plane glide reflection is also the restriction to the plane of an even isometry on space, namely a screw where the rotation is of 180°. "Clockwise" and "counterclockwise" are meaningless terms when applied to a plane in space without some particular reference system. This phenomenon carries forward to higher dimensions. A reflection on space is the restriction of a *rotation* in four-space. Transported into four-space, you would have to be careful not to drop your left shoe carelessly; your left shoe might get rotated and leave you with two right shoes.

The last sentence gives cause for a slight digression concerning chirality. In 1848 Louis Pasteur discovered that certain molecules exist in distinct right-handed and left-handed optically active forms depending on whether or not the molecules were from natural sources. This discovery has had far reaching consequences. We now know that although most compounds involved in fundamental life processes are chiral only one of the two enantiomorphic forms occurs in living organisms on Earth. Right-handed amino acids found in meteorites are known to be extraterrestrial because all amino acids originating on Earth are left-handed. All helices of protein and nucleic acids formed on this planet are right-handed. In four-space you would have to be careful your food didn't get rotated and become useless to you .The inorganic enantiomorphic forms of chiral substances on Earth are always found in equal proportions. Before 1957, physicists assumed there was no preference for one handedness over the other in the basic laws of physics. The contradiction of this assumption, which is called the "fall of parity," resulted from the celebrated work of Yang and Lee. In that the antinutrino is presumably the enantiomorphic form of the nutrino, the physicists have the first analogues on the particle level of Pasteur's enantiomorphic form of the tartaric acid found a

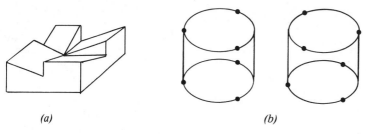

(a)   *(b)*

Figure 17.19

little over 100 years before in fermenting grapes. The existence and implications of chirality in space, time, and the elementary forces are at the forefront of modern research in physics. If antimatter is indeed the enantiomorphic form of matter, then the inconvenience of the two right shoes in the last paragraph is obliterated along with everything near the shoes as they annihilate each other. Apparently, having four dimensions of space would be hazardous to one's health.

Prisms and antiprisms, usually with "roofs" added to one or both bases, can be used to give polyhedra having certain symmetry groups. Figure 17.19a illustrates a roof that could be added when the base is a square. For our purposes, the essential difference between prisms and antiprisms based on regular n-gons is the following. For odd n the antiprisms have a point of symmetry but the prisms do not; for even n, the prisms have a point of symmetry but the antiprisms do not. The figures for these polyhedra are invariably too complicated to see what is going on, however. Therefore, these polyhedra are replaced in our figures by their circumscribed cylinders with added nodes corresponding to the vertices. The triangular prism and triangular antiprism are thus represented in Figure 17.19b. Further, the roof in Figure 17.19a will be represented by four arrows. For example, models having symmetry group C_n or D_n are illustrated in Figure 17.20 for $n = 3$ and $n = 4$.

$C_1, C_2, C_3, \ldots$ and $D_2, D_3, D_4, \ldots$ are the finite rotation groups we have so far. Are there more? The rotation groups of the Platonic solids have orders 12, 24, and 60, but none of these groups has a rotation of order greater than 5. Hence, none of these rotation groups is represented in the two lists above. In searching for all the finite rotation groups G, we shall consider three cases: (1) Group G contains only the identity and halfturns, (2) All the rotations in G of order greater than 2 have the same axis, and (3) Group G contains rotations of order greater than 2 with distinct axes.

The product $\sigma_n \sigma_m$ of two halfturns is a halfturn iff lines m and n are perpendicular. Hence, a rotation group containing only the identity and halfturns must have exactly 0, 1, or 3 halfturns. Therefore, a rotation group containing only the identity and halfturns is one of the cyclic groups C_1 or C_2 or else the dihedral group D_2. Group D_2 consists of the identity and the three halfturns about three mutually perpendicular axes. Case 1 has provided no new finite rotation groups.

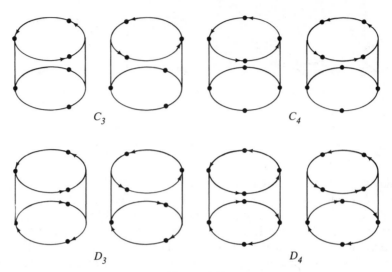

C_3 C_4

D_3 D_4

Figure 17.20

Next, suppose finite rotation group G contains rotations of order greater than 2 and that all these rotations have the same axis l. Since any rotation in G must fix l, the only possibility for rotations with axes different from l are half-turns about lines perpendicular to l. These perpendiculars must be concurrent since the group is finite. Therefore, if all the rotations of order greater than 2 in a finite rotation group have the same axis, the group is either C_n or D_n. Case 2 has provided no new finite rotation groups.

Since there is a point C such that each rotation in finite rotation group fixes C, the rotations in G are determined by their restriction to the unit sphere with center C. Each rotation in G has an axis that contains a diameter of the sphere and effects a *rotation* on the sphere about the points of intersection of the axis and the sphere. These points are called *poles* of the rotation. (Of course, if the rotation is viewed as clockwise about one pole, then the same rotation is viewed as counterclockwise about the antipodal pole.) Using the sphere helps in considering our third and last case, when G contains at least two rotations of order greater than 2 and with distinct axes. Since there are only a finite number of poles, we may suppose points P and Q are two nearest poles for rotations of order greater than 2. Let α and β in G generate the rotations in G with axes $\overleftrightarrow{CP}$ and $\overleftrightarrow{CQ}$, respectively. Suppose α and β have orders p and q, respectively. So $p \geq 3$ and $q \geq 3$. Then $\beta\alpha$ is a rotation in G, say of order r. See Figure 17.21a. A pole R of $\beta\alpha$ can be picked such that: the measure of the dihedral angle from the plane containing C, P, R to the plane containing C, P, Q has measure $360/2p$; the measure of the dihedral angle from the plane containing C, P, Q to the plane containing C, P, R has measure $360/2q$; and the measure of the dihedral angle from the plane containing C, P, R to the plane containing C, Q, R has measure $360/2r$. Since these dihedral angles have the same angle measure as the angles of the spherical triangle with vertices

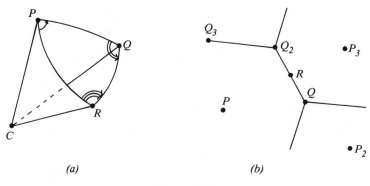

Figure 17.21

P, Q, R, then (Theorem 17.5) the sum of their measures is greater than 180. Hence,

$$\frac{1}{p} + \frac{1}{q} + \frac{1}{r} > 1.$$

Since $(1/p) + (1/q)$ is at most $2/3$, then $r < 3$. So $r = 2$. Since $r = 2$ and $q > 2$, then R is closer to P than Q is. Also, since $r = 2$, then $(1/p) + (1/q) > 1/2$ or, equivalently, $(p - 2)(q - 2) < 4$. We have five solutions:

(1) $p = 3, q = 3$;
(2) $p = 3, q = 4$;
(3) $p = 3, q = 5$;
(4) $p = 4, q = 3$;
(5) $p = 5, q = 3$.

Now, P is the center of a spherical regular p-gon with vertices the images of Q under the rotations with axis $\overleftrightarrow{CP}$. See Figure 17.21b. The rotations about Q move this p-gon to a set of q such p-gons surrounding Q, since P and Q are nearest poles of rotations of order greater than 2. Further, rotations of the same nature give a set of p-gons fitting together to cover the sphere. In other words, the images of Q form the vertices of a Platonic solid inscribed in the sphere. Similarly the images of P form the vertices of a dual Platonic solid inscribed in the sphere. There is no more room for rotations of order greater than 2 since P and Q are nearest poles of rotations of order greater than 2. The five solutions above do have realizations represented by the five Platonic solids. The images of $\overrightarrow{CR}$ are the axes of halfturns in G. We see that we have at least the rotation groups of the Platonic solids. Can there be more halfturns in G? Since poles P and Q are closest for rotations of order greater than 2, the only possibility for adding the pole of a halfturn would require a pole midway on the sphere between P and Q. However, such a halfturn composed with α would give a rotation of order 4 about $\overleftrightarrow{CR}$. Then R would be a pole of a rotation of order greater than 2 and nearer to P than Q is, a contradiction. There can be no more halfturns. Case 3 has provided only the groups $T, O,$ and

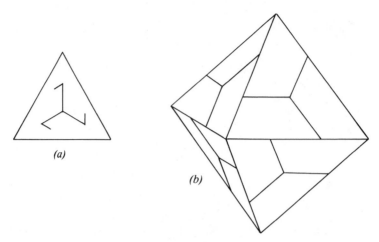

Figure 17.22

I to add to the cyclic groups C_n and the dihedral groups D_n. We have obtained all the finite rotation groups.

Theorem 17.10. *A finite rotation group on space is one of*

$$C_1, C_2, C_3, \ldots \qquad D_2, D_3, D_4, \ldots \qquad T, O, I.$$

Wire models having full symmetry group T, O, or I can be found by adding elements to the faces of the tetrahedron, the octahedron, and the icosahedron. These elements break any reflection or inversive symmetry but maintain all rotational symmetries. Each of these polyhedra has faces that are equilateral triangles, each triangular face is to be replaced by one having symmetry group C_3 such as Figure 17.22a. Another replacement that provides a wire model that "hangs together" is shown drawn on the faces of a solid octahedron in Figure 17.22b. You can see how you might add roofs on each of the trapezoids in the figure to give a polyhedron having full symmetry group O.

Now that we have all the finite groups of even isometries, we can begin to look for all the finite groups of isometries. We henceforth suppose G is a finite group of isometries containing an odd isometry. We may suppose (Theorem 17.9) that G fixes point C. Our search will entail two cases, distinguished by the presence or absence of σ_C in G. Half the elements in G are even and half the elements are odd. The even elements form a finite rotation group H. We have just determined all the possibilities for H. If G contains σ_C, then the odd isometries in G are the products of the elements in H with σ_C. Conversely, if H is a rotation group with elements $\alpha_1, \alpha_2, \ldots, \alpha_h$ fixing C, then these together with odd isometries $\sigma_C \alpha_1, \sigma_C \alpha_2, \ldots, \sigma_C \alpha_h$ form a set of $2h$ isometries denoted by $\overline{H}$. We claim H is a group. Since the axis of each rotation in H passes through C, then σ_C commutes with each α_i. So

$$\alpha_i(\sigma_C \alpha_j) = \sigma_C(\alpha_i \alpha_j), \qquad (\sigma_C \alpha_i)\alpha_j = \sigma_C(\alpha_i \alpha_j), \qquad (\sigma_C \alpha_i)(\sigma_C \alpha_j) = \alpha_i \alpha_j,$$

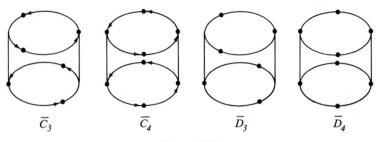

$$\overline{C}_3 \qquad \overline{C}_4 \qquad \overline{D}_3 \qquad \overline{D}_4$$

Figure 17.23

and $\overline{H}$ has the closure property. Also, $(\sigma_C \alpha_i)^{-1} = \alpha_i^{-1} \sigma_C^{-1} = \sigma_C \alpha_i^{-1}$ and $\overline{H}$ has the inverse property. Hence, for each rotation group H we have a new group $\overline{H}$. In addition to the rotation groups, we can now add

$$\overline{C}_1, \overline{C}_2, \overline{C}_3, \dots \qquad \overline{D}_2, \overline{D}_3, \overline{D}_4, \dots \qquad \overline{T}, \overline{O}, \overline{I}$$

to our list of finite groups of isometries.

Models for $\overline{C}_n$ and $\overline{D}_n$ are illustrated in Figure 17.23. To obtain these start with the antiprism models from Figure 17.20 if n is odd but the prism models if n is even and then add the required point symmetry. For D_n the arrows can be removed as the top and bottom have full symmetry. Since the octahedron (cube) and icosahedron (dodecahedron) have a point of symmetry, then $\overline{O}$ is the full symmetry group of the octahedron and $\overline{I}$ is the full symmetry group of the icosahedron. The tetrahedron, on the other hand, does not have a point of symmetry. Check that the wire model based on a cube in Figure 17.24 has full symmetry group $\overline{T}$. We know there must be more finite groups of isometries as we have not yet encountered the full symmetry group of the tetrahedron.

Returning to our search for all finite isometry groups, we suppose G is a finite group of isometries containing subgroup H of even isometries α_1, $\alpha_2, \dots, \alpha_h$ and containing odd isometries $\beta_1, \beta_2, \dots, \beta_h$. If some β_i is an inversion, then G is an $\overline{H}$. Hence, we suppose no β_i is an inversion. There is a point C that is fixed by G. With $\gamma_i = \sigma_C \beta_i$, the odd isometries in G are $\sigma_C \gamma_1$, $\sigma_C \gamma_2, \dots, \sigma_C \gamma_h$. Isometries $\gamma_1, \gamma_2, \dots, \gamma_h$ are distinct even isometries. Assuming $\alpha_i = \gamma_j$, we would have $\alpha_i = \sigma_C \beta_j$ and so $\sigma_C = \alpha_i \beta_j^{-1}$, a contradiction

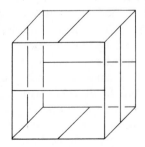

Figure 17.24

since σ_C is not in G. Hence, there are exactly $2h$ elements in the set K consisting of the even isometries $\alpha_1, \alpha_2, \ldots, \alpha_h$ and $\gamma_1, \gamma_2, \ldots, \gamma_h$. The surprising thing is that K turns out to be a group. Since (Exercise 17.1) an isometry fixing point C commutes with σ_C, then σ_C commutes with each element in G. Therefore, set K has the closure property and the inverse property since there are subscripts r, s, t, u such that

$$\gamma_j\alpha_i = \sigma_C\beta_j\alpha_i = \sigma_C\beta_r = \gamma_r,$$

$$\alpha_i\gamma_j = \alpha_i\sigma_C\beta_j = \sigma_C\alpha_i\beta_j = \sigma_C\beta_s = \gamma_s$$

$$\gamma_j\gamma_i = \sigma_C\beta_j\sigma_C\beta_i = \beta_j\beta_i = \alpha_t,$$

$$\gamma_i^{-1} = (\sigma_C\beta_i)^{-1} = \beta_i^{-1}\sigma_C = \beta_u\sigma_C = \sigma_C\beta_u = \gamma_u.$$

Group K is a rotation group since all the elements in K are even.

Conversely, suppose rotation group H has order h with elements α_1, $\alpha_2, \ldots, \alpha_h$, and suppose there is a rotation group K of order $2h$ having H as a subgroup and containing the additional rotations $\gamma_1, \gamma_2, \ldots, \gamma_h$ not in H. Let rotation group K fix point C. Then σ_C commutes with the elements in K. Let KH denote the set of $2h$ elements consisting of the h elements in H and the h odd isometries $\sigma_C\gamma_1, \sigma_C\gamma_2, \ldots, \sigma_C\gamma_h$. Nonempty set KH has the closure property and the inverse property since there are subscripts r, s, t, u such that

$$(\sigma_C\gamma_j)(\alpha_i) = \sigma_C(\gamma_j\alpha_i) = \sigma_C\gamma_r,$$

$$\alpha_i(\sigma_C\gamma_j) = \sigma_C(\alpha_i\gamma_j) = \sigma_C\gamma_s,$$

$$(\sigma_C\gamma_j)(\sigma_C\gamma_i) = \gamma_j\gamma_i = \alpha_t,$$

$$(\sigma_C\gamma_i)^{-1} = \gamma_i^{-1}\sigma_C = \sigma_C\gamma_i^{-1} = \sigma_C\gamma_u.$$

No inversion σ_D is in group KH as $\sigma_C\gamma_i = \sigma_D$ implies $\gamma_i = \sigma_C\sigma_D$, a contradiction (even if $C = D$). Hence, KH is a finite group of isometries containing odd isometries but no inversion.

In summary, the finite groups of isometries containing odd isometries are obtainable from the rotation groups in one of two ways. Starting with a rotation group H of order h that fixes point C, we obtain a group of order $2h$ with either method. The first method is simply to throw σ_C in with the group H to get new group $\bar{H}$. In $\bar{H}$, the odd isometries are just the products of the elements in H with σ_C. The second method is to find (if possible) a rotation group K containing H and with exactly $2h$ elements. The new group KH is formed by taking the elements in H and the odd isometries formed by multiplying the h elements of K that are not in H by σ_C where C is a fixed point of K.

The second method above is a recipe for finding all remaining finite groups of isometries. For the group H we begin with C_n. Of the finite rotation groups of order $2n$, group C_n is a subgroup of C_{2n} and of D_n. This gives new groups $C_{2n}C_n$ and D_nC_n. The only finite rotation group of order $4n$ of which D_n is a subgroup is D_{2n}. This gives the new groups $D_{2n}D_n$. The remaining possibilities for H are now limited to T, O, and I. Recall that these rotation

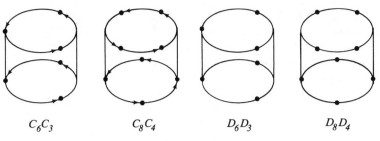

$$C_6C_3 \qquad C_8C_4 \qquad D_6D_3 \qquad D_8D_4$$

Figure 17.25

groups have orders 12, 24, and 60, respectively. There is no rotation group having half of its elements the 60 elements in I, and there is no rotation group having half its elements the 24 elements in O. However, there is one rotation group, namely O, that has half of its elements the 12 elements in T. Therefore, we have to add the group OT to finish our list of the finite groups of isometries. The group OT is the full symmetry group of the tetrahedron. We have proved *Hessel's Theorem*.

Theorem 17.11. *A finite group of isometries on space is one of*

$$C_n, \quad \overline{C}_n, \quad C_{2n}C_n, \qquad (n = 1, 2, \ldots)$$
$$D_n, \quad \overline{D}_n, \quad D_nC_n, \quad D_{2n}D_n, \quad (n = 2, 3, \ldots)$$
$$T, O, I, \quad \overline{T}, \overline{O}, \overline{I}, \quad OT.$$

For our cylindrical models, a rotation about the vertical axis through an angle of $(180/n)°$ is an even isometry (γ_1 in the argument presented above) in D_{2n} that is not in D_n. The roles of prism and antiprism are interchanged from the formulation of the models in Figure 17.23. Hence, Figure 17.25 illustrates models for $C_{2n}C_n$ and $D_{2n}D_n$. The odd isometries here include rotary reflections about the vertical axis. Halfturns about horizontal axes are rotations in D_n that are not in C_n. One of these composed with σ_C gives a reflection in a plane containing the vertical axis. The odd isometries in D_nC_n are n reflections in planes containing the vertical axis, as illustrated in Figure 17.26. Groups

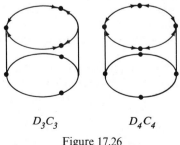

$$D_3C_3 \qquad D_4C_4$$

Figure 17.26

$D_n C_n$ are dihedral groups generated by a rotation and a reflection analogous to the dihedral groups encountered in studying plane isometries.

We have proved Hessel's Theorem and indicated figures having each of the possible finite symmetry groups. Every finite group of isometries is a symmetry group.

§17.3 Exercises

17.1. Show that an isometry fixing point C commutes with σ_C.

17.2. Show that a product of three rotations with axes that intersect in pairs but are not concurrent is a screw.

17.3. Describe the 24 odd symmetries of a cube.

17.4. Describe the 24 symmetries of a regular tetrahedron.

17.5. Excluding the regular tetrahedron, describe the symmetries of a pyramid having a regular n-gon for a base. Also, excluding the regular octahedron, describe the symmetries of the corresponding bipyramid.

17.6. Excluding the cube, describe the symmetries of a prism having a regular n-gon for a base. Also, excluding the regular octahedron, describe the symmetries of an antiprism having a regular n-gon for a base.

17.7. Describe the symmetries of a rhombohedron that is not a cube and the symmetries of a truncated octahedron.

17.8. Describe the symmetries of a square. Also, describe the symmetries of an orange, of a banana, and of a pear.

17.9. Prove or disprove: A group containing a rotary reflection that is not an inversion has infinite order.

17.10. True or False
(a) A finite group of even isometries is one of C_n, D_n, T, O, or I.
(b) Groups O and I are the only rotation groups G such that G is not a subgroup of a rotation group having twice the order of G.
(c) If $\sigma_x, \sigma_y, \sigma_z$ are the halfturns about the X-axis, Y-axis, and Z-axis, respectively, then $\sigma_y \sigma_x = \sigma_z$.
(d) If figures s_1 is a subset of figure s_2, then the symmetry group of s_2 contains the symmetry group of s_1.
(e) If a group G of isometries contains only the identity and rotations, then G is a cyclic group.
(f) Groups $C_{2n} C_n$, $D_n C_n$, and $D_{2n} D_n$ have order $2n$.
(g) Every subgroup of the icosahedral group is either I or else a group C_n or D_n.
(h) Group $C_2 C_1$ contains the identity and a reflection; group $\overline{C}_1$ contains the identity and an inversion.
(i) There is a unique point that is fixed by the elements of a finite rotation group.
(j) Distinct rotations α, β, and $\beta\alpha$ can have axes that are concurrent and coplaner.

17.11. Exhibit the elements in each of the following groups: $C_1, C_2, D_2, \overline{C}_1, \overline{C}_2, \overline{D}_2$, and $C_2 C_1$.

17.12. Make a paper model of a disphenoid from an isosceles triangle and another from a scalene triangle. Name the symmetry group for each disphenoid.

17.13. Name the symmetry group for the cuboctahedron and the symmetry group for the rhombic dodecahedron.

17.14. Name the symmetry group for the rhombicuboctahedron and the symmetry group for Sommerville's solid.

17.15. Name the symmetry groups associated with each of Exercises 17.4 through 17.8.

17.16. Describe the symmetries of a regular octahedron.

17.17. Describe the symmetries of a regular dodecahedron and the symmetries of a regular icosahedron.

17.18. Make enantiomorphic models of the snub cube. How many vertices, edges, and faces does a snub cube have? What is the symmetry group for the snub cube?

17.19. Make enantiomorphic models of the snub dodecahedron. How many vertices, edges, and faces does a snub dodecahedron have? What is the symmetry group for the snub dodecahedron?

17.20. If the pattern in Figure 17.27a is folded into a polyhedron, what is the symmetry group for that polyhedron?

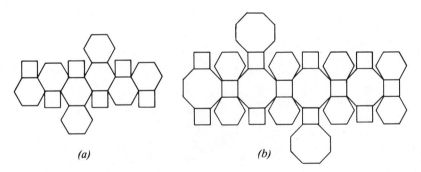

(a)   *(b)*

Figure 17.27

17.21. The pattern in Figure 17.27b can be folded into a polyhedron called a great rhombicuboctahedron. What is the symmetry group for this polyhedron?

17.22. Show that the product of two halfturns is a halfturn iff the axes of the halfturns are perpendicular.

17.23. Which Platonic solids cause the most problems in marking if the solids are tossed as "dice" in a game of chance?

17.24. Prove or disprove: There is a polyhedron with equivalent edges, with equivalent faces, but without equivalent vertices.

17.25. For each of the Platonic solids, describe the convex polyhedron determined by the union of the vertex figures of the Platonic solid.

17.26. Every finite group of isometries is a symmetry group. Is every group of isometries a symmetry group?

17.27. Use the concept of a centroid to show that a finite group of rotations must have a fixed point.

17.28. What are the finite groups of isometries that fix the set of all points in Cartesian space that have all coordinates integers?

17.29. A *deltahedron* is a polyhedron all of whose faces are quilateral triangles. Find all the convex deltahedra and give the symmetry group for each.

17.30. If a triangulation of a polygon in a plane results in t triangular faces, b vertices on the boundary of the polygon, and i vertices in the interior of the polygon, then show

$$t = 2i + b - 2.$$

Hints and Answers

Chapter 1

1.1. $\alpha, \delta, \eta, \sigma, \tau$,

1.2. δ: $3aX + 2bY + 6c = 0$, η: $bX + 3aY + 3(c - 2a) = 0$, σ: $aX + bY - c = 0$, τ: $aX + bY + (c - 2a + 3b) = 0$.

1.4. $Y = -5X + 7, Y = -5X - 7, Y = 5X - 7, X - 9Y - 32 = 0$.

1.5. Only g and j are false.

1.6. $x' = -x + 2a, y' = -y + 2b$.

1.12. $n^3, n(n + 1)^2, n^2(n + 1); n^2(n + 1), (n + 1)^3, n(n + 1)^2$.

Chapter 2

2.2. Multiply both sides by α^{-1} or β^{-1}.

2.3. Any two of α, β, γ determine the third if $\beta\alpha = \gamma$.

2.5. TTTFF TTFFF.

2.7. Rotation of 1 radian.

2.9. Consider groups of odd order.

Chapter 3

3.1. Given P, Q, R then $\sigma_R \sigma_Q \sigma_P$ fixes a vertex of the unique triangle.

3.2. A product of five halfturns that fixes Q.

3.3. Suppose bridge is $\overline{PQ}$. Let $\tau_{P, B}(Q) = R$. Opposite sides of a parallelogram are congruent. So $BP + PQ + QE = BR + RQ + QE$. BR is fixed length of bridge and $RQ + QE$ is minimum if equal to RE. Hence, the idea is to build the bridge $\overline{BR}$ first from B, determine point Q between R and E, and then translate the bridge to $\overline{PQ}$.

3.5. FFTTT TTFTT.

3.8. $x' = x + a_5, y' = y + b_5$.

3.11. $\sigma_D \sigma_C \sigma_B \sigma_A = \sigma_D \sigma_A \sigma_B \sigma_C = \sigma_B \sigma_A \sigma_D \sigma_C$.

3.18. The intersections of c_1 with $\tau_{C,D}(c_2)$ and $\tau_{D,C}(c_2)$ give the possible points A.

Chapter 4

4.2. 97 cm.

4.4. $C_2 = \langle \sigma_O \rangle, V$.

4.5. By unwinding from both ends, we see that one part of the path lies on the line through $\sigma_q \sigma_p(A)$ and $\sigma_p(E)$.

4.6. Only a and h are true.

4.10. Two letters.

4.12. You need to show α is a transformation.

4.14. Consider intersections of c_1 and $\sigma_l(c_2)$.

4.15. Think of the wall as a mirror.

Chapter 5

5.1. Perpendicular bisector of $\overline{AD}$ and $\overleftrightarrow{DE}$: $Y = -2X + 5, 4X - 3Y - 10 = 0$.

5.3. $\sigma_l \sigma_l = \iota$.

5.5. TTTFT FFFFF.

5.8. There are at most four and at least one, say α. Then, with the proper notation $\iota\alpha$, $\sigma_h\alpha$, $\sigma_v\alpha$, and $\sigma_O\alpha$ are the four.

5.9. If $Q = \sigma_m(P) = \sigma_n(P) \neq P$, then n and m are both perpendicular bisectors of $\overline{PQ}$.

5.10. Suppose directed angle is $\sqrt{2}°$.

5.14. With $P_Q = P_R$ and $a = \overleftrightarrow{RQ}$, then $a' = \overleftrightarrow{R'Q'}$.

5.15. D on $\rho_{A, \pm 90}(b)$.

Chapter 6

6.5. The perpendicular to m at $(0, 3)$ intersects n at $(2, 4)$: $x' = x + 2(2 - 0)$, $y' = y + 2(4 - 3)$.

6.6. $\sigma_l \sigma_m \sigma_l \sigma_l = \sigma_l \sigma_m$.

6.7. TFTTF TFTTT.

6.9. $\sigma_n \sigma_l \sigma_l \sigma_m = \sigma_n \sigma_m$.

6.11. Find two fixed points.

6.12. If $\rho(l) = l$ let $m \perp l$ and $\rho = \sigma_n \sigma_m$.

6.13. $\sigma_m \sigma_l \sigma_n$ is a reflection in q with $q \perp b$.

6.15. A reflection is an involution.

Chapter 7

7.3. $A^{\cdot} = B$, B is midpoint, B is on midline of parallel lines a and b, b is perpendicular bisector, A on b, b is midline or angle bisector.

7.6. $\rho_1 = \sigma_c\sigma_a$, $\rho_2 = \sigma_c\sigma_b$, $\rho_2^{-1}\rho_1$ fixes P.

7.7. TTTTF FTFFT.

7.9. Theorems 6.12, 6.11, 6.10.

7.12. Theorems 6.12. 6.6, 4.1.

7.13. Easy, $\rho_{D,\Phi} = \tau_{A,B}\rho_{c,-\Theta}$.

7.15. Given A, B, C, D want a, b, c, d. Take P such that $\rho_{P,90}(B) = C$. Let $E = \rho_{P,90}(D)$. $\overleftrightarrow{AE}$ is a.

Chapter 8

8.2. Follow the proof of Theorem 8.4.

8.3. Theorem 8.3.

8.5. Consider the easy case α odd first; then consider $\alpha\sigma_l$ when α is even.

8.7. TTFTF TFFTT.

8.12. If $\sigma_l\sigma_P$ where a dilatation δ, then σ_l would be the dilatation $\delta\sigma_P$.

8.17. $C_1 = \{\alpha_1, \ldots, \alpha_m\}$ and C_1 is a subgroup of order m in group G of order n. If β_2 in G but not in C_1, then $C_2 = \{\alpha_1\beta_2, \ldots, \alpha_m\beta_2\}$ has m elements and is disjoint with C_1. If β_3 in G but not in C_1 or C_2, then $C_3 = \{\alpha_1\beta_3, \ldots, \alpha_m\beta_3\} \cdots$. Eventually terminate with $C_1, C_2, \ldots, C_k$ and $n = km$.

Chapter 9

9a. Reflections and 1, 2, 3, 4, 5, ... ; two sixes back to back.

9b. 21 units.

9c. V is vertex fixed by σ_V where $\sigma_G\sigma_H(\sigma_F\sigma_B\sigma_E) = \sigma_G\sigma_H\sigma_I = \sigma_V$.

9l. Building two bridges from B and one from E is one possible beginning.

9p. Beware translations.

9s. $(0, 4)$.

9t. Theorems 8.3, 2.4, 3.4.

9w. $\rho_{C,60}$.

9x. With $m = \overleftrightarrow{AC}$, $n = \overleftrightarrow{AB}$, $Q' = \sigma_m(P)$, and $R' = \sigma_m(P)$, points R', R, Q, Q' are collinear.

9y. Since in 9x, $m\angle R'AQ' = 2m\angle BAC$ and $AR' = AP = AQ'$, then $R'Q'$ is minimal if AP is minimal; P, Q, R are the feet of the altitudes (vertices of the *orthic triangle*) of $\triangle ABC$.

9.4. $\alpha = (\alpha\sigma_l)\sigma_l$.

9.5. Involutory odd isometry, or odd isometry with fixed points.

9.6. Only b is false.

9.7. Look for fixed points.

9.8. Rotation about O through $(\Phi + 180)°$.

9.10. $2h = r - s/(\tan \Theta/2)$, $2k = s + r/(\tan \Theta/2)$, and $P = (h, k)$.

9.11. $(a - 1)X + bY + c = 0$; see 9.5.

9.12. Replace X, Y by x', y' to find the preimage of the line is the same line.

Chapter 10

10.1. $\mathscr{F}_1^3, \mathscr{F}_1, \mathscr{F}_2^1, \mathscr{F}_1^2, \mathscr{F}_2, \mathscr{F}_2^2, \mathscr{F}_1^1.$

10.5. TTFTF   TFFTT.

10.6. $\mathscr{F}_1, \mathscr{F}_1, \mathscr{F}_1^2, \mathscr{F}_1^1, \mathscr{F}_1, \mathscr{F}_1^2, \mathscr{F}_2^1, \mathscr{F}_2, \mathscr{F}_2^1, \mathscr{F}_2^1.$

10.8. C_1: 1984, C_2: 1961, D_2: 1881, D_1: 1883.

10.10. One is colorblind.

Chapter 11

11.1. $\overleftrightarrow{RS} = \sigma_T(\overleftrightarrow{PQ})$, $\sigma_{\overleftrightarrow{PS}} = \sigma_S \sigma_{\overleftrightarrow{RS}}$, $\overleftrightarrow{QR} = \sigma_T(\overleftrightarrow{PS})$, and Theorems 11.4 and 6.12.

11.2. Let $G_1 = \langle \rho_{A, 60}, \sigma_{\overleftrightarrow{CG}} \rangle$ and $G_2 = \langle \sigma_{\overleftrightarrow{AG}}, \sigma_{\overleftrightarrow{CG}}, \sigma_{\overleftrightarrow{AB}} \rangle$. Suppose $\rho_{A, 60}(B)$ $= C$, for orientation. Since $B = \sigma_{\overleftrightarrow{CG}}(A)$ and $\overleftrightarrow{BG} = \rho_{A, 60}(\overleftrightarrow{CG})$, then then $\rho_{B, 60}$ and $\sigma_{\overleftrightarrow{BG}}$ in G_1. So $G_1 = G_3$ where $G_3 = \langle \rho_{A, 60}, \sigma_{\overleftrightarrow{CG}}, \rho_{B, 60}, \sigma_{\overleftrightarrow{BG}}, \sigma_{\overleftrightarrow{AB}}, \sigma_{\overleftrightarrow{AG}} \rangle$. Likewise, show $G_2 = G_3$.

11.5. $\rho_{G, 120} \sigma_{\overleftrightarrow{AB}}, \rho_{G, 240} \sigma_{\overleftrightarrow{AB}}.$

11.8. $\gamma \varepsilon = \sigma_N$, $\varepsilon \gamma = \sigma_M$, $\gamma \sigma_A = \varepsilon.$

11.13. $\mathscr{W}_1^3, \mathscr{W}_2, \mathscr{W}_2^4.$

Chapter 12

12.1. Tiles: all except $3^4 \cdot 6$; edges: only $(3 \cdot 6)^2$.

12.3. If type p^q, then measures of interior angle $(360/q)$ and central angle $(360/p)$ add to 180 and so $(p - 2)(q - 2) = 4$.

12.7. TTTFT   TFFFT.

12.11. Prototile divides a hexagon having a point of symmetry with multiplicity 8.

12.12. Bow tie, leaf, and middle two of four heptominoes.

12.14. Those touching bottom edge of figure do not.

12.16. Cut $3 \cdot 4 \cdot 4 \cdot 3 \cdot 4$ into congruent infinite strips.

12.17. In $3 \cdot 4 \cdot 6 \cdot 4$ rotate by $30°$ some dodecagons formed by a hexagon and its adjacent squares and triangles.

12.24. Figure 11.33; F's cover Figure 10.12i two ways.

Chapter 13

13.1. A nonidentity stretch about C fixes exactly the rays with vertex C; a stretch rotation about C fixes no rays; a stretch reflection fixes exactly two rays.

13.3. $\alpha = \beta \delta_{P, r}$; $\delta_{P, r} \rho_{P, \Theta} \delta_{P, r}^{-1} = \rho_{P, \Theta}.$

13.5. $\sigma_m \delta_{G, 2}$ with G the centroid of $\triangle ABC$, G on m, and $m \parallel \overleftrightarrow{BC}$: G, m, $\overleftrightarrow{AG}$.

13.7. There is a similarity taking focus and latus rectum of one to focus and latus rectum of the other.

13.9. Only e and h are false.

13.18. Let L be the foot of the perpendicular from C to l. Let S be on $\overrightarrow{CA}$ such that $CS = 2CA$. Let T be on $\overrightarrow{CA}$ such that $CT = CB + CA$. Then M is on $\overleftrightarrow{CL}$ such that $\overline{SM} \parallel \overline{TL}$, and $m \perp \overleftrightarrow{CM}$ at M.

Chapter 14

14.1.   The dilatation that takes A to A' and B to B' takes $\overline{BC}$ to $\overline{B'C'}$.

14.2. $\overleftrightarrow{CF}$ is a transversal to $\triangle ABD$, and $\overleftrightarrow{BE}$ is a transversal to $\triangle ACD$.

14.5.   In Theorem 14.5 replace $\delta_{F,f}$ by $\tau_{A,B}$.

14.9.   Let p, q, r be lengths of perpendiculars from A, B, C to transversal $\overleftrightarrow{DE}$; $AF/FB = p/q$; for converse, let $\overleftrightarrow{DE}$ intersect $\overleftrightarrow{AB}$ at F'.

14.14. Start with any circle tangent to both rays.

14.15. First, find $\delta_{C,r}(P)$ for a point off the line.

14.18. If s is half the perimeter of the triangle, then $BL = s - AB$, $LC = s - AC$, $CM = s - BC$, etc.

14.20. I'.

14.22. $BP/CP = (AB/AC)^2$ when $\overleftrightarrow{AP}$ is tangent with B–C–P.

14.26. O and H are respectively the incenters of the tangential triangle and the orthic triangle.

Chapter 15

15.1.   What are the images of the perpendicular lines with equations $Y = X$ and $Y = -X$?

15.4. $15, \pm 15k, 15$.

15.6. $\alpha_h \alpha_k = \alpha_{hk}$ and $\beta_h \beta_k = \beta_{h+k}$.

15.7.   TFTTF   FTTTF.

15.9.   See proofs of Theorems 5.1 and 5.2.

15.14. $x' = x - y$, $y' = y$; $x' = x$, $y' = x + y$; only one fixed point.

15.17. $x' = 5x$, $y' = y/5$.

Chapter 16

16.1. $\sigma_D \sigma_C \sigma_B \sigma_A = \sigma_D \sigma_A \sigma_B \sigma_C = \sigma_B \sigma_A \sigma_D \sigma_C$.

16.4. $\sigma_\Pi \sigma_\Delta \sigma_\Gamma = \sigma_\Omega^{-1}$.

16.5. $\sigma_P \rho_{P,90} \delta_{P,2}$.

16.6.   TFFFT   TFTFT.

16.9. $\sigma_p = \sigma_\Pi \sigma_P$.

16.10. The group generated by the dilations.

16.12. First show $\alpha \sigma_l \alpha^{-1}$ is a halfturn.

16.21. $(4)(3) = 12$.

16.22. $(8)(3) = 24$.

Chapter 17

17.1. $\alpha\sigma_C\alpha^{-1} = \sigma_C$ iff $\alpha(C) = C$.

17.2.   Figure 17.18 and Theorem 17.6.

17.5.   Pyramid: rotations and reflections in $D_n C_n$. Bipyramid: symmetries in $D_{2n} D_n$ if n is odd but $\bar{D}_n$ if n is even.

17.6.   Prism: symmetries in $\bar{D}_n$ if n is even but $D_{2n} D_n$ if n is odd. Antiprism: symmetries in $D_{2n} D_n$ if n is even but $\bar{D}_n$ if n is odd.

17.10. TTTFF   FFTFF.

17.18. 24, 60, 38, O.

17.19. 60, 150, 92, I.

17.27. The centroid of the finite set of points consisting of all images of a given point must be fixed by each of the rotations since the set is itself fixed.

Notation Index

Index

Undergraduate Texts in Mathematics

Apostol: Introduction to Analytic
Number Theory.
1976. xii, 338 pages. 24 illus.

Bak/Newman: Complex Analysis.
1982. x, 244 pages. 69 illus.

Childs: A Concrete Introduction to
Higher Algebra.
1979. xiv, 338 pages. 8 illus.

Chung: Elementary Probability Theory
with Stochastic Processes.
1975. xvi, 325 pages. 36 illus.

Croom: Basic Concepts of Algebraic
Topology.
1978. x, 177 pages. 46 illus.

Fleming: Functions of Several Variables.
Second edition.
1977. xi, 411 pages. 96 illus.

Foulds: Optimization Techniques: An
Introduction.
1981. xii, 502 pages. 72 illus.

Franklin: Methods of Mathematical
Economics. Linear and Nonlinear
Programming, Fixed-Point Theorems.
1980. x, 297 pages. 38 illus.

Halmos: Finite-Dimensional Vector
Spaces. Second edition.
1974. viii, 200 pages.

Halmos: Naive Set Theory.
1974. vii, 104 pages.

Iooss/Joseph: Elementary Stability and
Bifurcation Theory.
1980. xv, 286 pages. 47 illus.

Kemeny/Snell: Finite Markov Chains.
1976. ix, 224 pages. 11 illus.

Lax/Burstein/Lax: Calculus with
Applications and Computing,
Volume 1.
1976. xi, 513 pages. 170 illus.

LeCuyer: College Mathematics with
A Programming Language.
1978. xii, 420 pages. 144 illus.

Macki/Strauss: Introduction to Optimal
Control Theory.
1981. xiii, 168 pages. 68 illus.

Malitz: Introduction to Mathematical
Logic: Set Theory - Computable
Functions - Model Theory.
1979. xii, 198 pages. 2 illus.

Martin: Transformation Geometry: An
Introduction to Symmetry.
1982. xii, 237 pages. 209 illus.

Millman/Parker: Geometry: A Metric
Approach with Models.
1981. viii, 355 pages. 259 illus.

Prenowitz/Jantosciak: Join Geometrics: A
Theory of Convex Set and Linear
Geometry.
1979. xxii, 534 pages. 404 illus.

Priestley: Calculus: An Historical
Approach.
1979. xvii, 448 pages. 335 illus.

Protter/Morrey: A First Course in Real
Analysis.
1977. xii, 507 pages. 135 illus.

Ross: Elementary Analysis: The Theory
of Calculus.
1980. viii, 264 pages. 34 illus.

Sigler: Algebra.
1976. xii, 419 pages. 27 illus.

Simmonds: A Brief on Tensor Analysis.
1982. xi, 92 pages. 28 illus.

Singer/Thorpe: Lecture Notes on
Elementary Topology and Geometry.
1976. viii, 232 pages. 109 illus.

Smith: Linear Algebra.
1978. vii, 280 pages. 21 illus.

Thorpe: Elementary Topics in
Differential Geometry.
1979. xvii. 253 pages. 126 illus.

Whyburn/Duda: Dynamic Topology.
1979. xiv, 338 pages. 20 illus.

Wilson: Much Ado About Calculus:
A Modern Treatment with Applications
Prepared for Use with the Computer.
1979. xvii, 788 pages. 145 illus.